ANIMAL BEHAV

GENES, DEVE
LEAR

ANIMAL BEHAVIOUR

A SERIES EDITED BY

T.R. HALLIDAY
Department of Biology
The Open University

AND

P.J.B. SLATER
School of Biology
University of Sussex

ANIMAL BEHAVIOUR · VOLUME 3

GENES, DEVELOPMENT AND LEARNING

EDITED BY T.R. HALLIDAY
AND P.J.B. SLATER

BLACKWELL SCIENTIFIC PUBLICATIONS
OXFORD LONDON EDINBURGH
BOSTON MELBOURNE

Editorial offices:
Osney Mead, Oxford OX2 0EL
8 John Street, London WC1N 2ES
9 Forrest Road, Edinburgh EH1 2QH
52 Beacon Street, Boston,
Massachusetts 02108, USA
99 Barry Street, Carlton,
Victoria 3053, Australia

First published 1983

Photoset by
Enset Ltd, Radstock Road,
Midsomer Norton, Bath, Avon
Printed and bound in
Great Britain by
Butler & Tanner Ltd,
Frome and London

Distributed in the USA and Canada by
W.H. Freeman & Co., San Francisco

British Library Cataloguing in
Publication Data

Genes, Development and Learning.—
(Animal behaviour; v. 3)
1. Animals, Habits and behavior of
I. Halliday, T.R. II. Slater, P.J.B.
III. Series
591.15 QH432

ISBN 0-632-00904-7
ISBN 0-632-00885-7 Pbk

CONTENTS

SERIES INTRODUCTION

As Niko Tinbergen, one of the founders of ethology, pointed out, if one asks why an animal behaves in a particular way, one could be seeking any one of four different kinds of answer. One could be asking about the evolutionary history of the behaviour: why did it evolve to be like it is? One could be asking about its current functions: through which of its consequences does natural selection act to keep it as it is? Thirdly, one might be interested in the stimuli and mechanisms that lead to the behaviour being performed: what causes it? Finally, one might be asking about development: how does the behaviour come to be as it is during the life of the individual animal? A complete understanding of behaviour involves investigation of all these questions, but in recent years there has been a tendency for ethologists to specialise in one or other of them. In particular, the functional analysis of behaviour has almost become a separate discipline, variously called behavioural ecology or sociobiology. This fragmentation of the subject is unfortunate, because all its facets are important and an integrated approach to them has much to offer.

Our approach in these books has been a more wide-ranging one than has been common in recent texts, with attention to all the kinds of explanation that have traditionally been the concern of ethologists. Aimed at students, each volume will provide a comprehensive and up-to-date review of a specific area of the subject in which there have been important and exciting recent developments. It is no longer easy for a single author to cover the whole field of animal behaviour with full justice to all its aspects. By asking specialists to write the chapters, we have tried to overcome this problem and ensure that recent developments in each area are fully and authoritatively covered. As editors, we have endeavoured to make sure that there is continuity between the chapters and that no significant gaps have been left in the coverage of the theme specific to each book. We hope that students who are inspired to further study by what they read will find the Selected Reading recommended at the end of each chapter a useful guide, as well as the more specific references which are gathered together at the end of each book.

We thank Bob Campbell and Simon Rallison of Blackwell Scientific Publications for their help and encouragement throughout the preparation of these books, Clare Little of Oxford Illustrators for her fine work on the illustrations and, most important of all, our authors for their readiness to accept a well-defined brief, to meet deadlines, and to accept our editorial changes and promptings.

T.R.H.

1983 P.J.B.S.

ACKNOWLEDGMENTS

Both of the editors and, in most cases, some of the authors of other chapters have commented on each chapter in draft. In addition, the authors of individual chapters would like to thank the following for their comments: Kevin Connolly, Richard Dawkins, Arthur Ewing and Aubrey Manning (Chapter 1); Robert Hinde and Joan Stevenson-Hinde (Chapter 2); Wolfgang Schleidt (Chapter 3); Jerry Hogan and Sara Shettleworth (Chapter 6). Marty Chalfie kindly allowed his unpublished findings to be quoted in Chapter 1.

INTRODUCTION

What influences shape behaviour during the lifetime of the individual animal? Of all the questions that can be asked about animal behaviour, those concerned with development have traditionally been the most contentious, partly because of differences in the emphasis of those looking at behaviour from different viewpoints. Ethologists, and especially those who worked on 'lower' animals, were impressed by the fixity of the behaviour that they studied, and by its remarkable adaptiveness. Trained as biologists, they appreciated that natural selection was responsible for this precision, and they tended to stress the genetic determinants of behaviour. 'Innate' behaviour was seen as arising fully formed the first time the animal required it and, at an extreme, development was sometimes seen as a subject for embryologists, of little relevance to behaviour (e.g. Lorenz 1965).

Diametrically opposed to this viewpoint was that of many psychologists, particularly those of the behaviorist school in America. They were interested in learning and intelligence, and in general laws which, though elucidated by work on animals, might be applicable to the study of humans. Their stress was thus on flexibility not fixity and on environmental not genetic determinants. To take an extreme view, J.B. Watson, the founding father of behaviorism, once wrote:

> 'Give me a dozen healthy infants, well-formed, and my own specified world to bring them up in, and I'll guarantee to take any one at random and train him to become any type of specialist I might select—doctor, lawyer, artist, merchant-chief and yes, even beggar-man and thief, regardless of his talents, penchants, tendencies, abilities, vocation and race of his ancestors.' (Watson 1930).

This was a strong claim, and Watson knew he was overstating his case, but he did so as an antidote to the fervour of those who believed that genes or 'breeding' were what really counted.

As must always happen with opposing viewpoints as extreme as these, it was just a matter of time before confrontation occurred, and this was followed by inevitable compromise. Confrontation came when a number of North American psychologists, notably Hebb, Lehrman and Schneirla, wrote criticisms of the emphasis placed by ethology, then largely a European science, on innate behaviour. Hebb (1953) had little sympathy for either extreme. He pointed out that to ask the extent to which a behaviour pattern was hereditary and the extent to which it was environmentally determined was:

> '. . . exactly like asking how much of the area of a field is due to its length, and how much to its width. The only reasonable answer is that the two proportions are one-hundred-per-cent environment, one-hundred-per-cent heredity. They are not additive; any bit of behaviour whatever is *fully* dependent on each.'

Such criticisms were well taken by most ethologists, and the more extreme forms of environmentalism also became less common amongst psychologists. A final round was fought between Lorenz (1965) and Lehrman (1970) but, by then, the whole dispute seemed to have tired itself out. The convention was to believe in a close and continuing interaction between genes and environment during the course of development. Ethologists had come to accept that environmental influences, including learning, might have a profound impact on the development of fixed and species-typical behaviour; for their part, psychologists came to appreciate the powerful way in which natural selection could lead to constraints on learning, making some animal species more capable of particular tasks than are others. The important task for both was to determine exactly *how* genes and environment interact with each other during the course of an animal's development.

Can one simply say, then, that the learning–instinct controversy is dead and that one need worry no more about it? Unfortunately this is far from being the case: the issues involved have a nasty habit of popping up elsewhere as soon as one thinks they are demolished, like the figures in a fairground shooting booth. It is certainly not a trivial matter; in everyday life the stress that people place on genes as opposed to environment in explaining their own behaviour and that of others has an immense impact on their attitude to life and to one another and is a major determinant of where they stand on the political spectrum. Amongst those whose writings may influence people, an extreme position on this issue, especially one which favours genetic determinism, may be positively harmful. Two recent examples illustrate this. One is the race and intelligence controversy, fired especially by an article by Jensen (1969) in which he argued that North American whites were intrinsically more intelligent than blacks. This was bound to be an unpopular viewpoint in that greatest of all environmentalist nations, but it was also logically unsound. Quite apart from the problems of measuring intelligence and the fact that tests devised to do so are likely to be more easily carried out by people of some cultural backgrounds than those of others, there is the problem of what is meant by heritability, this being the measure of inheritance used. It is not a measure of how much genes and environment each affect behaviour. As Hebb pointed out, that is a meaningless question. It is a measure of how much the *variance* in behaviour is attributable to each, originally devised for the selective breeding of plants; only where a good deal of the variance in a character was genetic could such selection be successful. The heritability measure is discussed further by Bateson in Chapter 2. Suffice it to point out here that the measure is as much affected by environmental differences as by genetic ones: heritability of a trait in one environment may be quite different from that in another.

The second example concerns sociobiology, that field of study on the border between ethology, ecology and evolution devoted especially to functional and evolutionary questions. Being concerned with the evolution of adaptive behaviour, and because natural selection can only work on genetic differences, this field is bound to stress the heritability of the behaviour patterns with which it is concerned. In some cases this point has, however, been overstressed in very much the same way as genetic influences on intelligence have been. This is curious, for while heritability may vary within wide limits, it is hard to conceive of any trait in which it would be negligible and which would not therefore be susceptible to change through natural selection. This, rather than strong genetic determination, is all that is needed for evolution to take place. Adaptiveness does not require behaviour to be genetically fixed. Yet E.O. Wilson, the leading figure in sociobiology, repeatedly stresses genetic determination as if behaviour was fixed and inflexible. To take an extreme example:

> 'Are human beings innately aggressive? This is a favourite question of college seminars and cocktail party conver-

sations, and one that raises emotion in political ideologues of all stripes. The answer to it is yes.' (Wilson 1978).

With such naive misunderstandings around, it is no wonder that others go to the opposite extreme and that those who study development sometimes despair! In this book, however, we are not grinding a genetic nor an environmental axe but trying, in a series of essays, to give something of the flavour of the field of behaviour development today. It is a field of immense importance, as we hope to have indicated in this introduction, both for our understanding of animal behaviour and how it comes to be as it is and also, more broadly, for various philosophical and political issues. Each of the six chapters is written by a different person. Although the last two chapters, those on learning, are the only ones written by trained psychologists, the impact of the recent dialogue between ethology and psychology will be apparent throughout. The assiduous reader will certainly find some points where chapters disagree, or where emphasis is different. This is inevitable in an active and fast-moving field, and we have left some overlap in coverage between chapters to allow such differences in perspective to persist. However, we trust that no reader will detect polemic. The emphasis is on an integrated and wide-ranging approach which will serve as both an introduction to, and a progress report on, what is one of the most exciting fields of behavioural research.

In the first chapter, Partridge describes recent developments in the genetics of behaviour. In the past, behaviour genetic analysis used often to consist of selection experiments to determine whether variance in a particular behaviour pattern had a genetic basis, and of a rather general comparison between animals known to be genetically different to see how their behaviour differed. But selection experiments nearly always work; as we pointed out above, it is hard to conceive of any behaviour pattern uninfluenced by varying genetic factors. Thus, these selection experiments were able to show, amongst other things, that the ability of rats to master mazes could be markedly raised or lowered by a few generations of selective breeding (e.g. Tryon 1940). Attempts to look at differences between genetically distinct animals came up against the problem that the relationship between genes and behaviour patterns is an exceedingly complex one, with each gene affecting many characteristics and each characteristic being influenced by many genes. Recently, however, as Partridge points out, new techniques have made possible more detailed study of how genes influence development. Two especially exciting examples that she discusses are the way in which it has become possible to study the precise effects of single-gene mutations on the simple nervous system of roundworms, and the use of mosaic individuals in fruit flies to pinpoint the structures responsible for the abnormal behaviour of mutants. Studies such as these, though often involving rather gross changes in the nervous system and behaviour patterns which are highly abnormal, are beginning to give an idea of the exact ways in which genes affect development.

Partridge also discusses the extent to which genes may be thought of as controlling development, but comes down firmly against attributing control to any particular one of the sources of information that affect the growing animal. She does not share Bateson's enthusiasm, expressed in Chapter 2, for culinary analogies and, as a result, the reader must choose whether or not to agree that development is like the baking of a cake. Suffice it to say here that no analogy should be taken too far, as both authors would, of course, agree. Waddington (1935) likened development to the Whitemoor marshalling yard of the London and North Eastern Railway, where trucks were rolled downhill through a series of points, the position of each of these progressively restricting the siding, or end-point, that they would reach. In some ways this was a good analogy with the narrowing of possibilities that accompanies ontogeny, with changes tending to occur at particular stages or sensitive periods. But, of course, as Bateson's discussion will make clear, the idea of tracks with no possibility of later transfer between them is altogether too limiting when we consider behavioural development. The cake analogy also breaks down if looked at in detail, but it neatly expresses the idea that the finished product in development is utterly dependent on a number of quite distinct factors whose contribution cannot be simply teased apart.

In Chapter 2, then, Bateson describes current views on how genes and environment interact to give behaviour. He shows just how far ethologists have progressed since the days when behaviour was thought to be either learnt or innate. But he also demonstrates that we have moved on from the opposite extreme, in many ways equally naive, that everything interacts with everything else in the course of development. There *are* rules which affect development (the existence of sensitive periods provides a striking example) and, indeed, development is to some degree regulated, so that some sudden gust cannot blow it permanently off course. Just what constraints there are on the outcome is a fascinating problem, and one with which the remaining chapters are all, to a greater or lesser extent, concerned.

Chapters 3 and 4 both consider the trajectories taken by growing animals: just what factors do affect the ways in which behaviour develops? Slater examines behaviour at the individual level and, in particular, whether or not experience of the situation to which it is adapted has a role in shaping it. In an unpredictable world, specific environmental inputs can help to gear the animal to the exact situation in which it finds itself. The development of the vertebrate visual system provides an example here, the genetic constraints on which were earlier discussed by Partridge. At the opposite extreme are behaviour patterns, of which anti-predator responses are the most obvious, which must be right the first time they are called upon. Experience of one sort or another can certainly affect their development, but it cannot be experience of encountering a hungry predator because that may have a lethal outcome. As in any discussion of behavioural development, bird song provides a key example of how genes and environment interact. It is referred to in several chapters but discussed in most detail by Slater. Although the learning of song from others is widespread, there are strong differences in strategy between species when one looks more closely; why this should be so is still a matter for speculation.

The way in which individual animals come to eat the right food, to avoid being eaten themselves or to sing an appropriate song is a complicated enough matter, but the development of relationships between animals is a good deal more so. Here it is not just how an animal adapts its behaviour to a largely indifferent environment, but how it modifies what it does in the light of the behaviour shown by others which may in turn be changing what they do in response to it. The development of relationships between animals is the subject covered by Chalmers in Chapter 4. The classic example here is that of imprinting, originally described as the process whereby young birds become attached to their mothers and may subsequently seek to mate with individuals which look like her. Recent research suggests, however, that one should differentiate between filial imprinting, involving the attachment to the mother, and sexual imprinting, which affects the choice of mate. The latter tends to take place rather later and to be affected by the young bird's experience of others as well as of the mother. Chalmers discusses imprinting in some detail, building on some of the points outlined by Bateson in Chapter 2.

The relationship between a young bird and its mother is a relatively simple case compared with those found in the social groups of many mammals. Here it is not simply a matter of the young animal following and maintaining contact with a single individual, but the infant must form relationships of different sorts with many individuals, and these may change as it grows older. Where two active participants are involved, it is not easy to work out who is responsible for changes which take place but, as Chalmers describes, there has been tremendous progress in this area, especially with experimental work on primates. This is an important topic, for mother–infant relationships in primates have similarities to those in humans and may help us to understand the reasons why these sometimes go awry.

The final two chapters are concerned with learning, a range of processes which have an important part to play in development and which continue, later in life, to enable adult animals to adapt their behaviour to changes in their surroundings. Mackintosh considers the great variety of different phenomena that we call learning. Some, such as sensitisation and habituation, are relatively simple and are found throughout the animal kingdom. Others, such as the more complex forms of associative learning that have been studied in detail in the laboratory by learning theorists, have been mainly described in higher animals. As Mackintosh points out, learning is certainly not a unitary phenomenon and it is, indeed, rather difficult to provide a definition of it which is unexceptionable. His chapter considers just what learning is and what forms it may take. While stressing diversity, he is at pains to point out that similar processes may be going on in a wide variety of different situations. One should not therefore throw up one's hands in horror at all attempts to generalise as, perhaps we might add, ethologists have recently been rather prone to do.

In the final chapter, Roper takes a lead from pointers Mackintosh provides to consider the biological importance of associative learning in greater depth and to make links between psychological and ethological work on learning. How has recent research, and especially that dealing with constraints on learning, left the extremely broad generalisations about learning that psychologists used to be so fond of making? Roper identifies two such generalisations: the principle of equipotentiality, which suggests that any animal should be able to learn anything within the limits of its sensory and motor equipment, and general process theory, suggesting that associative learning follows the same laws in all species. From this rather different perspective he approaches topics described in earlier chapters, such as imprinting and song learning, and asks what light the study of them has shed on learning and the psychologist's view of it. It is clear that learning must now be regarded as an adaptive phenomenon, a product of natural selection like any other attribute of an animal, and that, as such, broad generalisations about it can only be true in the loosest way. That this is so is perhaps most clearly illustrated by the changed view of learning that had to follow the discovery of learnt food aversion in rats. An animal which becomes ill several hours after eating a type of food will avoid eating it again: eating and illness are associated, but the time interval between them is far in excess of the close temporal proximity previously thought to be essential if learning was to take place. Examples such as this suggest that learning is best viewed as a collection of specialised abilities rather than as a single general process. Similarly the ethological literature is replete with examples to illustrate how far short the principle of equipotentiality falls as a general description of the abilities of animals.

One of the messages of this book is that the process of development is not as simple as it once seemed. The broad generalisations of early ethologists, and the different range of ones which psychologists put forward, were attractive and easy to grasp. But they were born out of ignorance at a time when few species had been studied in any detail. Some of these early ideas did, however, play a useful role as working hypotheses. Now that a substantial body of research has been conducted on behavioural development we have the information to see just where they fall short and we are able to replace them with more considered and firmly founded accounts of how development takes place. There is a long way to go yet to understand that exceptionally mysterious and refined process whereby a fertilised egg, having more in common with an amoeba than with a man, somehow gets translated into one of us. But, as far as behaviour is concerned, progress is being made and it is especially gratifying that the old entrenched attitudes of ethologists and psychologists have, as the chapters of this book will illustrate, been replaced by a greater understanding between them and by a feeling of common purpose.

CHAPTER 1 GENETICS AND BEHAVIOUR

LINDA PARTRIDGE

1.1 What is behaviour genetics?

It is now clear that variation in behaviour between species, populations and individuals often has some genetic basis, and that the genotype of an individual is therefore an important variable to be contended with in studies of behaviour (Manning 1975, 1976). While there is general agreement that genes can affect behaviour, people who study behaviour genetics may have a variety of questions in mind.

Sometimes it is the role of genetic variation in producing variation in behaviour that is the main interest. For example, an evolutionary biologist observing natural populations might want to know if variation in behaviour within a species has any genetic basis and, if so, how the genetic variation is maintained. A psychologist studying humans might be concerned with the role of genetic variation in producing variation in human behaviour, sometimes from an educational or medical standpoint. An agriculturalist or pest-controller might be interested in genetic variation in behaviour because of a wish to produce desirable behavioural characteristics in economically important animals.

In other cases, it may be the role of genes in producing normal behaviour, whether variable or not, that is being studied. The means by which genes act during the development of the nervous system and of the other structures which give rise to behaviour are very poorly understood. Genetic variation can be used to alter developmental processes in a way which helps us to understand how they are organised. The way in which behaviour is produced by the nervous system and other structures can also be studied by using genetic variation as a tool (Burnet & Connolly 1981). The neural basis of many aspects of behaviour, such as sensation, learning and sexual behaviour, have been studied in this way. Phenomena at a purely behavioural level can also be studied using genetic variants. For example, Willmund and Ewing (1982) have used a white-eyed mutant of *Drosophila melanogaster* to show that the red eye of the wild-type male is an important feature in eliciting female sexual receptivity.

Behaviour genetics, therefore, is a meeting-place of diverse interests and approaches and it is not possible to give a resumé of the whole area in a single chapter. Accordingly, no attempt is made to provide a comprehensive review of the subject. Several textbooks provide a broad coverage (Fuller & Thompson 1978; Hirsch 1967; Ehrman & Parsons 1981; Plomin *et al.* 1980). We shall consider some of the different approaches that have been used and, in particular, some of the very exciting new developments in the field. Since the publication of Manning's masterly accounts (1975, 1976), some important advances in behaviour genetics have occurred in the areas of neural development and the production of behaviour by the nervous system. In general the studies have involved the use of single-gene mutations to disrupt the nervous system, in the hope that a comparison between mutant and normal individuals can throw light on the way the nervous system and behaviour are organised in genetically normal individuals. These studies are beginning to give us some idea of *how* genes exert their effects on behaviour. To an extent this approach marks a break with some more traditional studies which have sometimes ignored the processes and structures involved in mediating the effects of genes on behaviour. This new phase in one aspect of behaviour genetics is still in its infancy, and we are very far from complete understanding of how any gene exerts its effects on behaviour, but the general approach of studying genetic effects at the level of the nervous system seems a very promising way forward. The first part of the chapter is devoted to some examples of such studies. This is followed by a discussion of some behavioural traits that are affected by genetic variation at many gene loci, and lastly some recent evolutionary studies of behaviour genetics are described.

1.2 The use of single-gene mutants to study the nervous system and behaviour

1.2.1 *The basic approach*

Behaviour and the structures producing it can be studied using both observational methods and experimental disruption. Several different techniques can be used to observe and disrupt behaviour at various levels. Neurophysiological methods can be used to record naturally occurring nerve impulses and to stimulate nerves electrically. Hormone levels can be monitored and manipulated. Anatomical studies and surgery can be used to observe, remove or alter the position of particular structures. Behaviour may also be studied by simply watching the animal, and then manipulated by varying certain external stimuli or the animal's internal state.

Single-gene mutants can also be used as a tool for studying behaviour. Mutations useful for this sometimes arise spontaneously but more often are induced by mutagens, especially the chemical ethylmethanesulphonate (EMS). In a mutant individual, all the cells of the body carry the mutant gene, but mutations can sometimes have very specific effects, altering only one or a few aspects of structure and behaviour. Sometimes these effects cannot be produced using other techniques. A comparison between normal and mutant individuals can throw light on the way the nervous system and behaviour are organised in normal individuals.

In this section, fairly detailed acounts are given of some studies where mutants have proved useful in the study of the nervous system and behaviour. Two general features of mutants should be borne in mind; these arise because so-called pleiotropic mutations can have many different effects on the animal, often because one abnormality leads to another during development. First, a mutation could produce its effect on behaviour in several ways. For example, the cricket *Achaeta* has a pair of sensory appendages, the cerci, on the rear of the abdomen. These cerci bear wind-sensitive hairs, innervated by sensory neurons the axons of which project to the abdominal ganglion where they synapse with the medial giant interneuron (MGI) of the ventral nerve cord. Bentley (1975) isolated and studied a wind-insensitive cricket mutant. The MGIs in these mutant crickets had stunted dendrites that were thinner than those in normal crickets. This might suggest that the abnormal MGI morphology was responsible for wind-insensitivity. However, the wind-sensitive hairs were also absent in the mutant, providing an additional or alternative explanation for wind-insensitivity. The second important point about mutants is that where a mutation has more than one effect, the effects may not be independent. For example, one possible explanation for the abnormal MGI morphology of the mutant cricket is that the mutation has two separate effects, one on the wind-sensitive hairs, the other on the MGI dendrites. Alternatively, the absence of hairs may mean that the sensory neuron does not carry the normal nerve impulses to the MGI, and normal development of the MGI dendrites may require a normal electrical input from the sensory neuron. This second interpretation is supported by the observation that a very similar stunted anatomy of MGI dendrites is found in genetically normal crickets that have developed with their cerci surgically removed (Murphey *et al.* 1975). It is often the case that a particular neural abnormality is the result of a 'cascade of pleiotropic effects' initiated by a (usually unknown) primary effect of the mutation and it is important to bear this in mind (Stent 1981).

The case of the cricket cerci clearly illustrates that studies of mutants are at their most convincing when they are combined with the use of other techniques. Only then is it possible to be sure that the sequence of events is fully understood. Ideally, if we wish to combine the analysis of mutants with other types of analysis, we require an organism with a good repertoire of behaviour patterns, a relatively small number of large nerve cells (to facilitate identification of individual cells and neurophysiology) and well-understood genetics with a large array of mutants (Quinn & Gould 1979). Unfortunately, such an organism does not exist and all the organisms studied so far have both advantages and disadvantages.

1.2.2 *Touch-sensitive cells in nematodes*

The development of the nematode *Caenorhabditis elegans* has been very extensively studied over the last decade.

The animal has several advantages for a genetic study. It is transparent throughout its life cycle, so that direct observations can be made on individual cells and the patterns of cell division in the living worm. This has led to the discovery that *C. elegans* develops in an almost invariant manner, with precise spatial patterns of cell division. These divisions give rise to particular families of cells, known as cell lineages (Sulston & Horvitz 1977; Deppe *et al.* 1978; Kimble & Hirsch 1979). These cell lineages in turn give rise to adults with a reliable number and arrangement of somatic cells. In addition, the animal is small; the adult hermaphrodite (*C. elegans* occurs as hermaphrodites and males) is about 1mm in length and has 959 somatic cells, 302 of which are neurons. The small size of the animal has allowed the anatomy of the nervous system to be completely reconstructed from serial section electron micrographs (see, for example, Ward *et al.* 1975; White *et al.* 1976; Sulston *et al.* 1980). The number, position, cell morphology and cell lineage history of the nerve cells in *C. elegans* are essentially invariant in wild-type individuals. Because of its short life cycle (3½ days), *C. elegans* is well suited to genetic analysis; its genetics are well understood (Brenner 1974; Herman & Horvitz 1980) and an impressive array of mutations has been induced and characterised. Many mutations with an effect on behaviour have been found. For example, mutants with abnormal locomotion (Brenner 1974), chemotaxis (Ward 1973, 1977; Dusenbury *et al.* 1975; Dusenbury 1980; Lewis & Hodgkin 1977) and thermotaxis (Hedgecock & Russell 1975) have been isolated.

A detailed study has been made of touch-insensitive mutants (Chalfie & Sulston 1981; Chalfie & Thomson in press; Chalfie in press and pers. comm.). These mutations are very specific in their effects; the touch insensitive mutant worms can move normally and respond to a strong mechanical disturbance, but will not move in response to a gentle touch from an eyebrow hair. The mutants have proved very useful in studying the role of genes in the development and functioning of the touch receptors of *C. elegans*.

The touch receptors in *C. elegans* are a set of six neurons, the mechanosensory cells (Chalfie & Thomson 1979). These cells (see Fig. 1.1 for their arrangement and names) all possess an anteriorly directed process that runs close to the body cuticle, separated from the cuticle only by a thin layer of hypodermis. In the electron microscope the cells can be identified by their unusually large microtubules, and also by a structure called the mantle between the cell process and the hypodermis.

Several mutants have been isolated in which touch sensitivity is abolished. The mutations involved occur at 12 different loci and

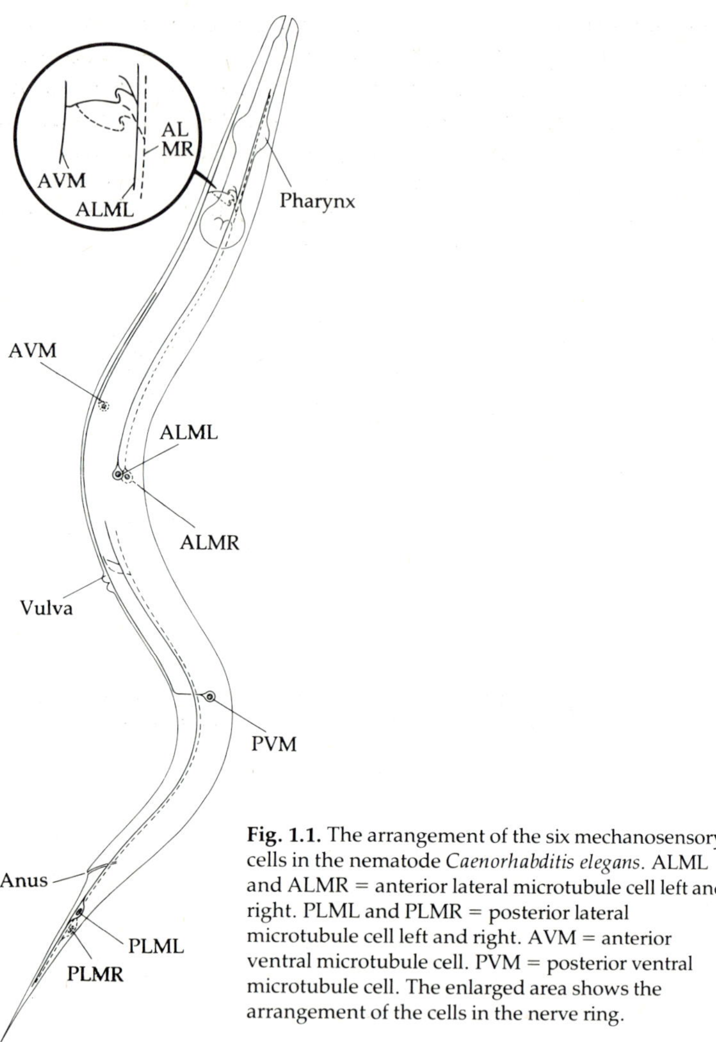

Fig. 1.1. The arrangement of the six mechanosensory cells in the nematode *Caenorhabditis elegans*. ALML and ALMR = anterior lateral microtubule cell left and right. PLML and PLMR = posterior lateral microtubule cell left and right. AVM = anterior ventral microtubule cell. PVM = posterior ventral microtubule cell. The enlarged area shows the arrangement of the cells in the nerve ring.

seven of them have visible effects on the touch cells (Chalfie & Sulston 1981; Chalfie pers. comm.).

In some touch-insensitive mutants, the outgrowth of the touch cells is normal, but their function is impaired (Fig. 1.2b and c). This has allowed the identification of cell components necessary for normal cell function. For example, one of these mutants (*mec-7*) does not contain the large microtubules usually found in the mechanosensory cells, and these microtubules may be necessary for sensory transduction. In another mutant (called *mec-1*), the mechanosensory cells have no mantle, and the cell processes are not properly embedded in the hypodermis. It has been suggested (Chalfie & Sulston 1981) that the mantle may act as glue, causing the mechanosensory-cell process to stick to the hypodermis, and this may be necessary for normal sensory transduction to occur.

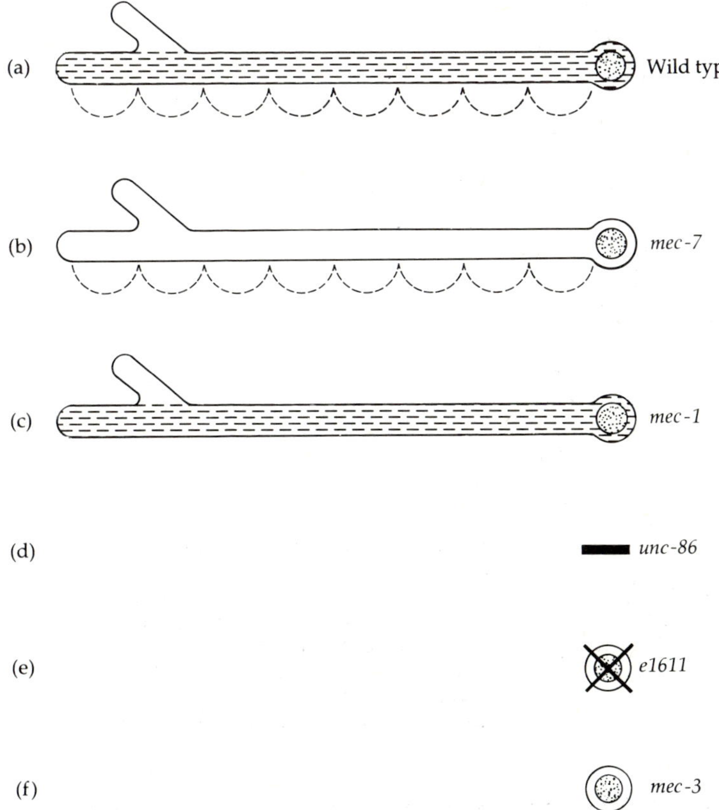

Fig. 1.2. Diagram of the effect of various mutants on the mechanosensory cells of *C. elegans*. (a) Wild type with cell body, receptor process, large microtubules and mantle; (b) *mec-7*, with no large microtubules; (c) *mec-1*, with no mantle; (d) *unc 86*, with no mechanosensory cells; (e) *e1611*, in which the mechanosensory cells degenerate after the cell divisions which give rise to them are complete; (f) *mec-3*, which lacks the receptor process.

Another group of mutations (Fig. 1.2d–f) causes abnormal growth or development of the mechanosensory cells. For example, *unc-86* alters the patterns of cell division which give rise to the cells, so that they are never produced, while another mutation, *e1611*, causes the cells to degenerate after they are produced. Worms displaying the *mec-3* mutation lack the receptor process and this mutant gene must block further growth of the cells during development. It is interesting that such a small number of genes seems to be specifically necessary for these stages of growth to occur; during the original process of mutagenesis, multiple copies of these mutations were isolated, suggesting that all the genes specifically necessary for normal mechanosensory-cell growth have been identified.

One mutation, *mab-5*, produces worms with normal touch sensitivity but has an effect which suggests that in nematodes some cells use cues derived from the position they occupy in the body in determining their pattern of differentiation. The *mab-5* mutant gene has the effect of putting the PVM cell in the wrong place; its precursor cell migrates further forward than usual, so that its process terminates in an anterior position usually only occupied by the AVM cell process. In normal worms the cell processes of the AVM and PVM cells have different morphologies. The PVM cell process is unbranched and has a gap junction with the AVM cell process. The AVM cell process has a branch, the synaptic branch, making synapses with other neurons in the nerve ring, the major area of neuropil in *C. elegans*. Interestingly, in *mab-5* worms the anteriorly displaced PVM cell process develops a similar synaptic branch. This suggests that the synaptic branch is induced by the position of the cell process relative to other structures, because a cell (the PVM) in an abnormal position develops structures usually only found in cells in that position. Such an interpretation is also suggested by the finding that, in genetically normal worms, laser lesions which prevent the AVM precursor cell from making its normal movement forwards during development also result in an AVM cell with no functional synaptic branch. By contrast, some aspects of the mechanosensory-cell morphology seem to be unaffected by abnormal position. For example, the microtubules and the mantle are not affected by *mab-5*, and the length of the PVM receptor process also seems to be normal.

The development of nematodes had previously been viewed as

very determinate, and for the most part laser ablation experiments have confirmed this since, on the whole, developing nematode cells seem to be little affected by the removal of adjacent cells. Recently, however, some examples have been found where cell position does seem to affect development (Sulston & White 1980; Kimble 1981a,b; Kimble & White 1981). Altered cell position can have a number of effects, including changes in cell lineage, change in the cell types produced and alterations in cellular structure. The *mab-5* mutant has provided a good example of the importance of cell interactions in the developing nervous system, and has also provided a means of teasing apart those aspects of cell morphology which are affected by cell position and those which are not.

As yet, the exact way in which each of these mutations exerts its effect is not known. However, as more becomes known about the genes involved, their products (if any) and the timing of their activities, we should be able to see a much clearer picture of the roles of specific genes in mechanosensory-cell development and function. Of course, these cells are only involved in one aspect of sensation; there will still be a very long way to go to understand the roles of genes in the development of the whole worm nervous system with its complicated pattern of synaptic connections. Nonetheless, the study of the mechanosensory cells has certainly made a start in this direction.

1.2.3 The development of the cat visual system

Unlike nematodes, mammals have extremely complicated nervous systems containing millions of cells and they also show far more elaborate patterns of behaviour. It seems inevitable that understanding of the role of genes in the development of the more simple nervous systems will come first. Nonetheless, mutants are already providing very valuable insights into the processes at work in the developing mammalian nervous system.

In the Siamese cat, and some other mammals with reduced pigmentation, there are abnormalities in the visual pathway which result in a partially scrambled input to the visual centres in the brain. The way in which the brain handles these abnormal inputs has provided information about the way in which an ordered system of connections between the retina and the brain may be

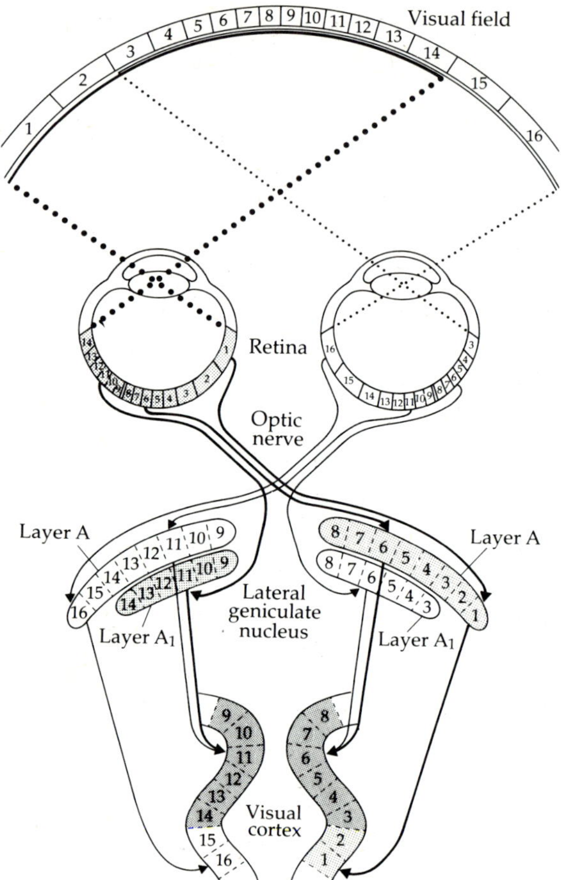

Fig. 1.3. The neural connections in a genetically normal cat. The visual field is divided into 16 unequal-sized segments. Each half of the brain 'looks' at the half of the visual field on the opposite side, and relatively more brain tissue represents the central than the peripheral part of the visual field. An inverted image of the visual field falls on the retina, and the left eye does not see the extreme right (segments 15 and 16), nor the right eye the extreme left (segments 1 and 2), of the visual field. Nerve fibres from the nasal part of each retina then cross over (decussate) to project to the lateral geniculate nucleus on the opposite side, while fibres from the temporal part of the retina project to the lateral geniculate nucleus on the same side. Representations of the parts of the visual field seen by both eyes (segments 3–14) are localised in two layers (A and A_1) in the lateral geniculate nucleus: projections from the eye on the same side go to A_1 while those from the eye on the opposite side go to A. The representations in the lateral geniculate layers are in register so that 'lines of projection' (broken lines) through the layers represent single points in the visual field. The two sets of representations are then fused in the cortex.

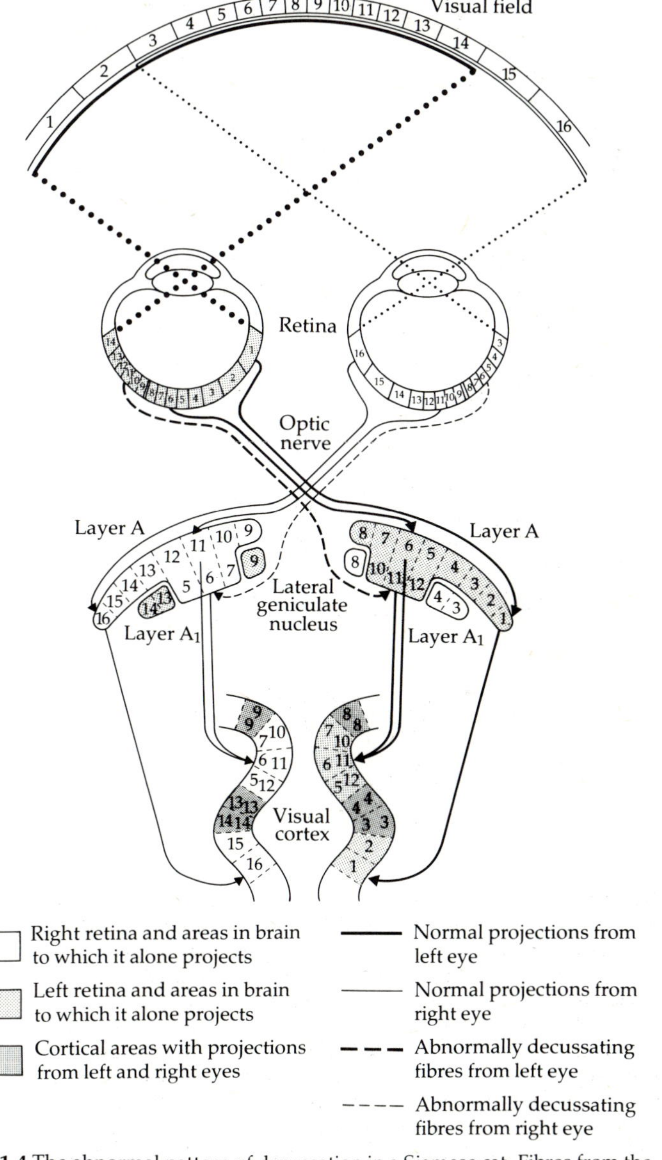

Fig. 1.4 The abnormal pattern of decussation in a Siamese cat. Fibres from the more central areas of the temporal retina decussate, resulting in abnormal projections in layer A_1 of the lateral geniculate nucleus. If the normal pattern of projections to the cortex were then followed, projections from the two layers of the lateral geniculate nucleus instead of reinforcing each other would in many cases conflict, with neural activity reflecting stimuli at two different points in the visual field (e.g. 6 and 11). This is not the pattern of cortical projections actually seen in Siamese cats (see Fig. 1.5).

achieved during normal development. The abnormality of the visual pathway has been best studied in Siamese cats (for reviews see Guillery *et al.* 1974; Guillery 1974; Stent 1981; Lund 1978; and Gaze & Keating 1972).

In mammals, visual information is relayed from the retina via the lateral geniculate nucleus (LGN) to the cerebral cortex. It is in the LGN that the inputs from the two eyes are matched before being passed on to the cortex. Some retinal ganglion cell axons from each eye cross over, or decussate, at the optic chiasm to innervate the contralateral (opposite side) LGN. The pattern of decussation in normal cats is shown in Fig. 1.3; fibres from the nasal half of each retina decussate while those from the temporal half do not. The temporal part of the retina is that nearest the side of the head and the nasal part is, as its name suggests, that nearest the nose. In cats, as in other carnivores and primates, each eye sees nearly all of the visual field, there being a small area on the nasal side of each retina which sees a part of the visual field invisible to the other eye. Each half of the brain 'looks' at the opposite half of the visual field because of the pattern of decussation.

When the fibres from the retina reach the LGN they synapse with two cell layers. Fibres from the ipsilateral (same side) eye, projecting from the temporal part of the retina, synapse with the inner cell layer (A_1), while fibres from the nasal part of the contralateral retina synapse with the outer cell layer (A). Each LGN layer thus receives a map of one visual hemifield, and the maps in the layers A and A_1 are in register so that a line roughly perpendicular to the layers corresponds to a single point in the visual field.

In the Siamese cat, this pattern of neural projections from the retina to the LGN is disrupted because of an abnormal pattern of decussation (Fig. 1.4). Some temporal nerve fibres, which would normally all go to the ipsilateral LGN, instead decussate. Once these misdirected fibres arrive at the LGN they seem to behave normally in that (a) they are fibres from the temporal part of the retina and they synapse with layer A_1 in the LGN, and (b) fibres from more medial positions on the retina go to more medial parts of the LGN and fibres from more temporal retinal positions go to more temporal parts of the LGN.

Thus the abnormally decussating fibres go to the wrong LGN, but when they arrive there they form a normal set of projections. However, this results in abnormal visual information reaching the

LGN. First, within layer A_1 different sections of the visual field are sometimes discontinuous or left–right inverted with respect to each other. When the cat moves its head laterally the inverted segments cause the LGN to receive conflicting information about the direction of movement. Secondly, the layers A and A_1 are in some places out of register so that some columns in the LGN receive input from different parts of the visual field.

These observations suggest two important conclusions. First, during development the system of innervation of the LGN layers is guided mainly by the retinal position of origin of the axons. Which eye the fibres come from, and the visual coherence of the information they produce, do not seem to affect the innervation pattern. Secondly, the mechanisms governing decussation of the nerve fibres must be different from those by which the same fibres are caused to connect with particular parts of the LGN. In the Siamese cat, decussation is abnormal but the abnormally routed fibres make normal contacts with the LGN.

Are the projections from the LGN to the visual cortex affected? In a normal cat (see Fig. 1.3) most cortical cells receive inputs from the same areas of both layers of the ipsilateral LGN and therefore have a binocular input and can be activated through either eye. There is an orderly point-to-point projection from the LGN to the cortex, so that the visual fields are mapped on to the cortex.

If this wiring pattern were followed in Siamese cats, some cortical loci would receive a contradictory input; neural activity at one cortical locus would represent visual stimuli in two different parts of the visual field. In addition, as in the LGN layer A_1, movement would be represented in a contradictory way. All this would make it very difficult for a Siamese cat to see and the cortex is in fact wired differently in these animals. Two different patterns of connections have been found (Fig. 1.5).

In the first, all the information from the abnormal LGN layer A_1 is suppressed, even though in some parts of the LGN the two layers are in normal register. The suppression of these normal segments shows that lateral interactions must play some part in determining the input to the cortex. Suppression of the information from layer A_1 means that the animal cannot see with the temporal part of the retina, and this is confirmed by behavioural observations on these cats.

In the second pattern seen in Siamese cats (Fig. 1.5), there is a

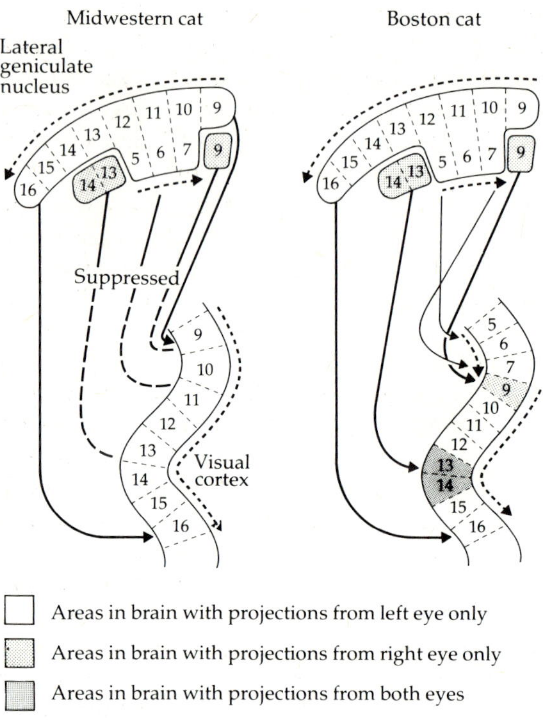

Fig. 1.5. The two patterns of connections actually found between lateral geniculate nucleus and cortex in Siamese cats. In the Midwestern cat, the input to the cortex from layer A_1 is largely suppressed, eliminating the information from abnormally arranged projections 5–7. In the Boston cat, the projections from 5–7 are instead reversed and inserted at a different point in the cortex, so that a more or less continuous picture of the outside world is produced.

complete reorganisation of the projections from LGN to cortex. Both the sites and the ordering of the projections are abnormal, in such a way that the visual field segments are represented in order within each side of the cortex. The sequence which was disrupted in the LGN is therefore corrected during development of the cortex. This re-ordering of the projections has the consequence that each side of the cortex 'sees' a larger segment of the visual field than in a normal cat. The visual field of each eye is normal, however.

Both of these abnormal patterns allow the cortex to receive an unambiguous and sequentially organised picture of the visual fields, but neither allows binocular vision. It is not clear what determines which of these two solutions is adopted, although there is some evidence that severe cross-eyes, an abnormality often seen in Siamese cats, is associated with the second pattern.

These observations suggest that the innervation of the visual cortex cannot be guided mainly by the position of origin of the LGN cell axons but rather demands visual coherence of the information coming from axons making functional connections. The cortex will not accept a disrupted representation of the visual field. How the incoming axons are monitored and either re-routed or suppressed is not understood.

Thus the genetic abnormality in Siamese cats has given us great insight into the adaptive plasticity found in the developing cat visual system; it is this plasticity which is undoubtedly responsible for the fine tuning of binocular vision during normal development (Blakemore 1977; see also section 3.2). What does the Siamese cat tell us about the role of genes during development?

Why is it that Siamese cats show the abnormal pattern of decussation? They carry a temperature-sensitive allele of the gene which makes tryosinase, an enzyme which catalyses one stage in the synthesis of the pigment melanin. In Siamese cats the pigment is only made at temperatures below 37°C, the body temperature of cats, so that only the extremities of the animal are pigmented. Impaired decussation seems to be a general property of albino mammals and it has been found in tigers, mice, rats, ferrets and mink. Some of these albinos lack, not tyrosinase, but some other enzyme involved in melanin synthesis, suggesting that it is lack of pigment rather than lack of tyrosinase which is the crucial variable.

Why should lack of melanin cause abnormal decussation? There is no melanin in the retinal ganglion cells whose axons show the abnormal decussation pattern. However, underlying the retina is a layer of melanin-containing cells, the pigment epithelium. There is some evidence that it is the presence of pigment in this epithelium which is important. Some mammals, such as Burmese cats and chinchilla rabbits, are deficient in pigment in the skin but their visual system shows the normal wiring pattern. They also have normal pigment epithelium. In mink there is a series of mutations which affect the pigmentation of various parts of the body, and in this series the extent of abnormal decussation is inversely related to the amount of pigment in the pigment epithelium of the eye. During development, differentiation of the retinal ganglion cells occurs next to the pigment epithelium layer after melanin has been laid down. We therefore have strong circumstantial evidence that a lack of pigment in the epithelium somehow produces abnormal decussation, but the mechanism by which it does so is at present a mystery.

This example has been treated at some length because it highlights some strengths and weaknesses of the use of mutants to study nervous system development. On one hand, the abnormal decussation seen in albino mammals would have been extremely difficult to produce by other methods, such as surgery, and it has produced some fascinating information about visual system projections. On the other hand, the nature of the primary defect which causes abnormal decussation is not understood, whereas with surgical intervention the nature of the lesion is usually known. In the case of the Siamese cat, therefore, the mutant has so far told us quite a lot about the ways in which interactions between nerve cells can affect development, and rather little about the role of genes in the process. However, once we understand *how* the abnormal decussation is produced, we will be much closer to understanding normal decussation and the role of genes therein.

1.2.4 Drosophila *learning*

Neurophysiological analysis of the processes involved in learning is difficult because the smallest isolated parts of the nervous system capable of associative learning still contain many neurons. However, there have been some notable advances in the study of the neural basis of associative learning in molluscs such as *Pleurobranchaea* using the techniques of neurophysiology (Davis 1976). Recently, analysis of mutants has been used as a tool in the study of associative learning in *Drosophila*. The approach is still a fairly new one but some interesting mutants have been isolated.

Associative learning was first demonstrated in *Drosophila* using an olfactory learning task, where either of two different odours was paired with an electric shock (Quinn *et al.* 1974). After three 15-second training periods, normal flies remember for 3–6 hours and selectively avoid whichever of the two odours was associated with the shock during training. In similar experiments, flies learned to avoid light of a particular wavelength (Quinn *et al.* 1974). *Drosophila* associative learning has since been demonstrated in other situations (Menne & Spatz 1977; Siegel & Hall 1979). Some of these studies have the disadvantage that they were done with populations of flies, and the measures of learning used were based on the difference in the proportion of the population approaching the conditioned stimulus before and after conditioning. This technique does not rule out the possibility that only a very few flies may in fact learn and the others may simply follow them or not show any change in behaviour. Associative learning has been demonstrated using single flies (Medioni & Vaysse 1975; Siegel & Hall 1979), so that this may not be a serious problem, but it is worth pointing out that the value of any genetical analysis is limited by the validity and reliability of the phenotype measure.

Using the olfactory test developed by Quinn *et al.* (1974), two learning-deficient mutants have been isolated (Dudai *et al.* 1976; see also section 2.6). These two mutations are two alleles of the same gene, and are named *dunce*. The *dunce* flies do not learn in the olfactory task; after training they do not selectively avoid the shock-associated odour. As well as the adults, *dunce* larvae are also deficient in olfactory learning (Aceves-Piña & Quinn 1979). However, *dunce* individuals seem to show normal behaviour in other respects, and the effects of *dunce* do not apply to all types of learning because *dunce* flies show normal learning in a visual task (Dudai & Bicker 1978). Using a slightly different olfactory learning task, Dudai (1979) demonstrated that *dunce* flies can form a short-lived (less than 15 minutes) memory of training, and this suggests that *dunce* flies may be blocked in an early stage of memory formation.

The *dunce* mutation has been found to map to the same chromosome position as independently isolated mutations causing female sterility and deficiency in the enzyme cyclic AMP (adenosine monophosphate) phosphodiesterase. Flies of *dunce* phenotype have an abnormally low level of this enzyme, as do individuals displaying the female-sterile mutations mapping to the same locus; the latter mutants learn poorly, like *dunce*. This suggests that all three mutant phenotypes are produced by an abnormality of the same gene (Byers 1980; Dudai & Quinn 1980).

Cyclic AMP phosphodiesterase breaks down cyclic AMP, so that *dunce* flies have an abnormally high level of cyclic AMP.

Furthermore, normal flies display poor learning after being fed on phosphodiesterase inhibitors, suggesting that cyclic AMP phosphodiesterase is involved in, or necessary for, olfactory learning in *Drosophila*. Changes in the level of neuronal cyclic AMP have been implicated in the process of sensitisation in *Aplysia* (Klein & Kandel 1978), and the finding that it may also be involved in associative learning in *Drosophila* is interesting, although perhaps not very surprising.

Very recently, using a modified version of the olfactory learning task, another mutant with defective memory has been isolated (Quinn *et al.* 1979). In this case, the mutant flies learn normally but forget abnormally rapidly, although not as rapidly as *dunce* flies, and the mutant has been named *amnesiac*. So far, no further analysis on this mutant has been published.

1.3 The use of mosaics to locate the structures producing mutant behaviour

Analysis of mosaic animals has been used to locate the anatomical site(s) at which a mutation exerts its effect(s). Mosaics are composite individuals in which some tissues are mutant and the rest are wild type at a particular locus. By studying many different mosaics it is sometimes possible to identify which body parts must be mutant if mutant behaviour is to be shown. This elegant and powerful technique has been used particularly in *Drosophila* and in mice.

It is important to realise that mosaic analysis does *not* identify all the structures necessary to produce a particular behaviour. It identifies the structures which must be mutant to produce mutant *as opposed to wild-type* behaviour. Throughout this section, when it is stated that a particular structure must be mutant for a particular mutant behaviour to be produced, the expression 'mutant behaviour' is being used as a shorthand for 'mutant as opposed to wild-type behaviour'.

1.3.1 Techniques for making Drosophila *mosaics*

There are two main methods for producing *Drosophila* mosaics (reviewed by Hall *et al.* 1976).

One method relies on the induction of mitotic recombination

during development. A heterozygous embryo or larva, with a recessive behavioural mutant on one chromosome and the dominant wild-type allele on the homologue, is exposed to X-rays at some point during development. This is done in order to induce crossing-over between homologous chromosomes which, after mitosis is complete, may give rise to a daughter cell homozygous for the behavioural mutation. This cell divides normally, giving rise to a clone of mutant cells, and the later in development the animal is exposed to X-rays, the smaller is the clone of mutant cells produced in the adult.

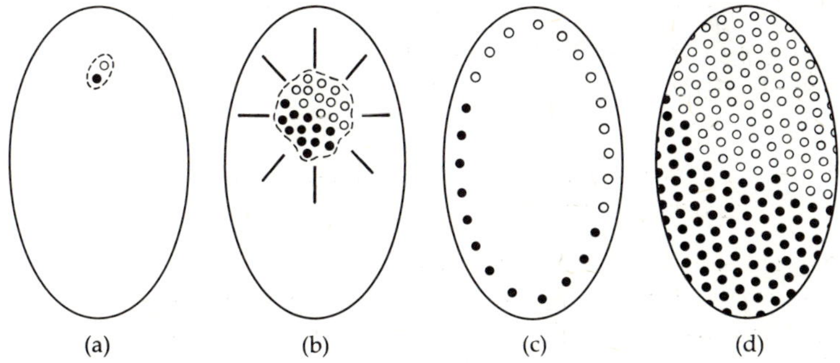

Fig. 1.6. The formation of a mosaic fly by the loss of the ring chromosome (X_R). X_R is lost in about 35% of XX_R eggs at first division of the zygote nucleus, producing (a) one XX_R nucleus (solid circle) and one XO nucleus (open circle). Nuclei divide in a cluster (b) and then migrate to the egg surface (c). (d) shows this surface as seen from outside; the mosaic boundary separates two areas, one XX_R and wild type, one XO expressing the behavioural mutant and X-linked tissue-markers. These two areas give rise to different adult structures, so that the adult fly is also a mosaic.

In the second technique, much more commonly used in behavioural analysis, gynandromorphs (sex mosaics) are produced (see Fig. 1.6). To do this, a female zygote (genotype XX) is produced in which one of the X chromosomes is ring shaped and unstable. There is a high probability of the ring X being lost from a nucleus during one of the very early nuclear divisions in development, giving rise to a clone of XO genotype (XO in *Drosophila* is male, although wild-type males are XY). Thus a clone of male cells is produced in an originally female embryo. The position of the boundary between male and female tissue is very variable, depending upon the orientation of the first few nuclear divisions relative to the eventual body axis. If, in the original female zygote, the normal X chromosome carried a recessive behavioural mutation, then the gynandromorph will also be mosaic for the mutant phenotype, male tissues being mutant and female tissues being wild type (see Fig. 1.7).

Fig. 1.7. Typical *Drosophila melanogaster* mosaics. Shaded areas are XX_R and female. Unshaded areas are male and express recessive mutant characters uncovered by the loss of the ring X.

One technical problem with mosaic analysis is that some means is needed to identify which cells or tissues are mutant and which are not. In *Drosophila* mosaics this is done by having recessive tissue-markers genetically linked to the recessive behavioural mutation. Cells with the mutant behavioural phenotype, therefore, also have the mutant tissue-marker phenotype.

The tissue-markers can be used in two ways. The first uses mutations which alter the surface structures of the fly, in conjunction with a technique called fate mapping (Hotta & Benzer 1970, 1972), to pinpoint the internal tissue(s) responsible for mutant behaviour. Subsequent work has made mosaic analysis more precise by using the second technique, in which the internal tissues are marked directly. One method uses genetic variants of the enzyme acid phosphatase to mark mutant and normal tissues (Kankel & Hall 1976). Recently, a new marker has been developed using variants of the enzyme succinate dehydrogenase (Lawrence 1981). With both these techniques, staining and serial sectioning of the fly are necessary to score the internal tissues.

Hotta and Benzer (1972) mapped the foci of mutant activity of several behavioural mutants by using their fate mapping technique. For example, a mutant called *drop dead*, causing premature death in adult flies, mapped to the brain, while another called *wings up* mapped to the thoracic muscles. The indirect flight muscles of *wings up* flies were subsequently found to be histologically abnormal or even absent.

In these cases, the mosaic technique was used to indicate where in the fly to look for a lesion. The method identifies the *primary* focus of the mutation. It might have been quite simple without mosaic analysis to discover that *wings up* was associated with abnormal indirect flight muscles, but the mutation's primary effect could have been on the nerves innervating these muscles, leading to their atrophy or abnormality. Fate mapping placed the primary action of *wings up* in the muscle itself.

1.3.3 *Circadian-rhythm mutants in* Drosophila

Mosaic analysis has been used in the study of circadian-rhythm mutants in *Drosophila*. Rhythmic variations in behaviour have been found in many organisms, and when the rhythm persists under constant conditions and has an approximately 24 h period it is called a circadian rhythm.

Three circadian-rhythm mutants were induced in *Drosophila* by Konopka and Benzer (1971). These were an arrhythmic mutant (per^{o}) showing no circadian rhythm in the timing of emergence from the pupa (eclosion) or in activity, a short-period mutant (per^{s}) which had an approximately 19 h rhythm, and a long-period mutant (per^{l}) which had an approximately 28 h rhythm. These three mutations have subsequently been found to have analogous effects on short-term rhythms in courtship song (Kyriacou & Hall 1980).

Genetic mapping suggested that all three mutations were in the same functional gene, and mosaic analysis indicated that the rhythm phenotype corresponded to the genotype of the head (Konopka & Benzer 1971). Brain transplants were then performed, in which the brains of per^{s} flies were transferred to the abdomens of *per*^{o} (arrhythmic) flies (Handler & Konopka 1979). Of 55 individuals surviving the operation, four showed short-period activity rhythms for at least three consecutive cycles. Degenerative changes in the implanted brains may have been responsible for the 51 negative results. Because the transplanted brains had no functional neural connections with the host nervous system, it seemed very likely that some humoral substance was involved. Anatomical studies on the brains of *per*^{o} flies in both *D. melanogaster* and *D. pseudoobscura* have shown that the position of some neurosecretory cells is often abnormal in *per*^{o} mutants, although the range of anatomy in mutants shows considerable overlap with that of wild-type flies (Konopka & Wells 1980). The significance of the abnormal position of these cells is not clear and similar information on the position of the cells in the *per*^{s} and *per*^{l} flies has not been published. *Per*^{o} flies have a 75% reduction in octopamine synthesis in the brain (Livingstone 1981), but whether this is related to the abnormal position of the neurosecretory cells is not known.

1.3.4 *Gynandromorphs and courtship in* Drosophila

Drosophila courtship behaviour has been extensively analysed using gynandromorphs to establish what parts of a fly must be male or female for particular sexual behaviours to be shown.

Hotta and Benzer (1976) used body surface markers and fate mapping to determine which parts of a fly had to be male for it to show the following and wing-vibration elements of male courtship (Fig. 1.8). The focus mapped to the area of the embryo which usually gives rise to the brain. Only flies which followed and vibrated their wings showed attempted copulation, the focus for which mapped to the thoracic ganglia. These conclusions were confirmed and extended by Hall (1977, 1979), who used the internal tissue-marker acid phosphatase as well as cuticle markers. Hall concluded that male tissue in the left or right side of the dorsal brain was necessary for male following and wing vibration, and in addition some male tissue in the thoracic ganglia was necessary for attempted copulation. Interestingly, the courtship behaviour of gynandromorphs was less vigorous than that of control males, so it is probably necessary for tissues other than the brain and thoracic ganglia to be male in order to produce normal male courtship.

Gynandromorph analysis has also been used to investigate courtship song. During male wing vibration in *D. melanogaster* two sorts of courtship song are produced, pulse song and sine

Fig. 1.8. Mosaic flies. Female cuticle is dark, male cuticle is light. The 40 flies on the left were chosen randomly from the group that did *not* show wing vibration, while the 40 on the right were chosen randomly from the group that did show wing vibration.

song. Working with gynandromorphs, Schilcher and Hall (1973) showed that in order for pulse song to be produced, the mesothoracic ganglion had to be male. The other song type, sine song, could not be correlated with male tissue at any single site.

The gynandromorph results indicate that in *Drosophila* the genotype of individual cells, rather than the level of circulating hormones, is responsible for the sexual phenotype.

1.3.5 *Techniques for producing mouse mosaics*

A genetically mosaic mouse may be constructed by removing early embryos (or morulae) of two different genotypes from their mother's oviducts, dissociating them by enzyme treatment and mixing the cells together. The cell mixture is then incubated for some hours and the mixed cells reassociate, continue to cleave, and form a single blastocyst which is then reimplanted into the uterus of a female host animal. There the mosaic blastocyst continues normal embryonic development and gives rise to a mosaic mouse.

In these mosaic mice the proportions of the total cell population derived from the original two morulae varies within wide limits. Some mosaic individuals are equal mixtures of the two cell types, while in others as few as 1% of the cells may be of one type. Others show no evidence of mosaicism. This variability results partly from the fact that only a few of the morula cells contribute descendents to the mouse, many contributing only to extra-embryonic structures. Thus a large variety of mosaic patterns can be generated.

In these mouse mosaics the two cell types are much more finely mixed than in *Drosophila* mosaics. In the latter, whole limbs or larger areas of the body are often of one genotype, but in mouse mosaics cells of a single genotype occur in much smaller patches (Mullen 1978).

As with *Drosophila* mosaics, some means is needed to recognise which cells in the mosaic carry a behavioural mutation. To do this, the two morulae, only one of which carries the behavioural mutation, are also made to carry different genetic variants of an enzyme which can be scored in individual cells in stained histological sections. These mouse mosaics can then be used in a similar way to *Drosophila* mosaics, namely to pinpoint the primary site of action of a mutant gene.

A second technique used in constructing mouse mosaics relies on the fact that in female mice only one of the two X chromosomes is active in each cell in the body. If the two X chromosomes in a single individual carry different alleles, then different body cells will express the alleles of only one X chromosome, so that the mouse will be phenotypically mosaic.

1.3.6 *Mutations affecting the mouse cerebellum*

Mutations causing defective locomotion are easy to detect, and it is perhaps for this reason that several such mutations have been found in mice. Most of these affect the structure of the cerebellum, the part of the brain controlling the execution of fine movements.

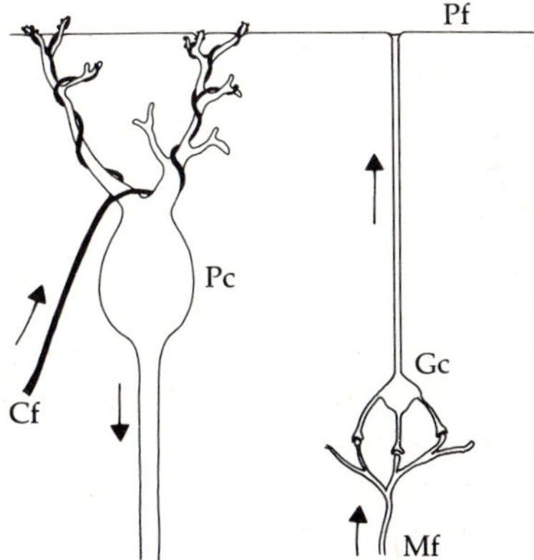

Fig. 1.9. The wiring pattern of the cerebellum in genetically normal mice. Pc = Purkinje cell, Mf = mossy fibre, Cf = climbing fibre, Pf = parallel fibre, Gc = granule cell.

In normal mice the cerebellum contains a number of different cell types, arranged in a characteristic pattern (Fig. 1.9). The fibres constituting the neural input to the cerebellum form two distinct channels. The climbing fibres project directly to the Purkinje cells, while the mossy fibres terminate on the granule cells, which each have a T-shaped axon, the parallel fibre, synapsing with the Purkinje cells.

In one mutant, called *staggerer,* locomotion is grossly abnormal and the animal has tremors. The cerebellar granule cells are absent. During development in a genetically normal mouse, the granule cells are formed on the outer surface of the cerebellum and migrate inwards to their eventual position. In *staggerer* mice, the granule cells are formed normally and make their inward migration. They at first form a normal set of projections to the Purkinje cells, but subsequently the granule cells degenerate and disappear, as do the mossy fibres.

In *staggerer* mice the Purkinje cells look abnormal; they do not have the normal spiny branchlets on the end of their dendrites and they are greatly reduced in number (Caviness & Rakic 1978; Sotelo 1980). Mosaic analysis has been used to investigate whether the granule cells themselves are defective in *staggerer* mice or whether the lesion is in some other cell type such as the Purkinje cells.

In *staggerer*/normal mosaic mice made using the cell mixing technique, the cerebellum had a morphology intermediate between normal and mutant (Mullen & Herrup 1979). The granule cell layer was nearly normal, but the Purkinje cell layer was a patchwork of cells with normal position and dendrites and cells with abnormal positions and spineless dendrites. The abnormal Purkinje cells were of *staggerer* genotype, while the normal Purkinje cells were wild type. This suggests that the *staggerer* genotype was having a direct effect on the Purkinje cell morphology. Unfortunately, the granule cells were too small for the staining technique to be used to identify their genotype, but on other grounds it is highly likely that the granule cells were a genetic mosaic arranged independently from the Purkinje cell mosaic. Thus, although there is room for doubt, the *staggerer* mutant probably has a primary effect on the Purkinje cells with the granule cells being secondarily affected. The results from the mosaic mice also suggest that, provided some Purkinje cells are genetically normal, the granule cells are also normal, which might be related to the fact that the granule cell axons form synapses with several different Purkinje cells, some of which are likely to be genetically normal. Thus some normal postsynaptic (Purkinje) cells are probably needed to maintain the presence of the presynaptic (granule) cells. The exact role of the Purkinje cell dendritic spines is at present not clear.

Another mouse mutant, *reeler*, has abnormalities in both the cerebellum and the cerebral cortex. In the cerebellum, the Purkinje cells fail to move to their normal position during development. Mosaic analysis has shown that the genotype of the Purkinje cells does not determine their pattern of migration; some Purkinje cells migrate normally in mosaics, but normal migration is unrelated to *reeler*/wild-type genotype. The fact that some Purkinje cells do migrate normally in the mosaics suggests that some other group of cells may help guide the Purkinje cells to their usual destination, and that these other cells may be affected in *reeler* mice. In mosaics the migration of each Purkinje cell may be normal if it encounters a genetically normal cell in the population of 'guiding' cells which is postulated.

1.3.7 *Mutants affecting sexual behaviour in mice*

The X-inactivation technique (see section 1.3.5) has been used to examine how much of the mouse central nervous system must be male for adequate male sexual behaviour to be shown.

In mice, male sexual behaviour is only shown in individuals which were exposed to androgen shortly after birth. The sex-linked testicular feminisation (*Tfm*) mutant causes androgen insensitivity and prevents this neonatal determination. Another sex-linked mutation, sex-reversal (*Sxr*), acts early in development and causes XX individuals to become morphologically male. Ohno *et al.* (1974) have constructed heterozygous mice with two X chromosomes, one bearing *Tfm* and *Sxr*, the other bearing the wild-type alleles at these two loci. These mice are morphologically male because they carry *Sxr*, and they are mosaic for androgen-sensitivity because in some cells *Tfm* is expressed while in others it is not, as a result of the pattern of X-inactivation. This means that during the neonatal determination process, cells expressing *Tfm* are not imprinted while cells expressing the wild-type allele are imprinted and therefore have the usual male phenotype.

In normal XX mice the pattern of X-inactivation is random, each X chromosome being inactivated in approximately half the cells in the mouse. In mice there is an X-linked locus, *O*, at which the allele O^{hv} alters the pattern of inactivation in O^{hv}/O^{+} heterozygotes, with approximately 4:1 odds favouring the preferential activation of the X chromosome carrying the O^{hv} allele. Ohno *et al.* used this locus in conjunction with *Tfm* and *Sxr* to produce mice in which X-inactivation meant that only one cell in four was phenotypically male. Somewhat to their surprise, Ohno *et al.* found that some of these mice showed normal male levels of aggressiveness towards an intruding male and normal male behaviour towards an oestrous female. Thus a mouse does not need anything like 100% fully functioning androgen-sensitive neurons to display adequate male behaviour; there must be considerable redundancy at the neuronal level in these animals.

1.3.8 Conclusions

Mosaic analysis is still a relatively new technique which has not yet been fully exploited. Nonetheless, some interesting findings are already appearing, although it is probably fair to say that few of the *Drosophila* results to date are very surprising. As mosaic analysis is refined and extended to more species, its value will undoubtedly increase. In *Drosophila* the rather approximate technique of fate mapping will probably be to some extent replaced by the more accurate internal marking techniques. The sort of subtle disruption seen in *staggerer*/wild-type mosaic mice would be extremely difficult to produce in any other way, and it has cast considerable light on the way normal neural connections may be formed in the mouse cerebellum.

1.4 Quantitative variation and the study of behaviour

1.4.1 Quantitative characters

So far, cases have been considered where animals can be classified into discrete phenotypic classes between which there are no intermediates. All the mutants discussed show behaviours or neural characteristics which are distinctly different from those shown by normal, wild-type members of the same species. This sort of discontinuous variation has many advantages for further analysis; it allows many experimental animals with predictable characteristics to be produced, it makes the scoring of phenotypes reasonably simple and it allows a straightforward analysis of the results. Single-gene mutations often have pleiotropic effects; one of the first examples of a gene mutation with an effect on behaviour was the *yellow* mutation in *Drosophila,* which affects sexual behaviour as well as eye colour (Bastock 1956), and we have seen other examples such as those producing reduced pigmentation in mammals (see section 1.2.3). Thus a mutant gene can affect several characters. We shall consider now the converse situation, namely cases where a single character is affected by genetic variation at many gene loci. This is a common finding with naturally occurring variation. Much of the naturally occurring variation in behaviour, and in the structures producing it, is continuous; for many phenotypic characters a population cannot be classified into discrete categories between which there are no intermediates. Much variation of this type is environmental in origin, but genetic variation may also be involved and, where it is, many gene loci are often found to affect variation in a single behaviour pattern. These continuously varying characters which may be affected by variation at many gene loci are often called quantitative characters and the study of the genetic variation involved is called quantitative or biometrical genetics (Falconer 1981; Mather & Jinks 1971). This type of genetic variation has been used, like single-gene mutants, as a tool in the study of behavioural organisation.

1.4.2 Quantitative genetics and behavioural organisation

The usual approach when studying behavioural organisation by using quantitative genetic variation is to produce lines of animals which show a difference in the mean value of some behavioural trait. This is usually done either by inbreeding or by selection.

Inbreeding is the mating of individuals which are related by ancestry (Falconer 1981). If close relatives, such as a brother and sister, are mated together and a pair of their progeny are then mated together, and so on for many generations, the offspring become increasingly homozygous with each generation. This happens because during inbreeding the two parents will, by chance, be identical homozygotes for one or more loci. Subsequently, heterozygosity at these loci can only be restored in the inbred line by mutation. Eventually, an inbred line is produced, the members of which all have the same completely homozygous genotype. For each locus which was genetically variable in the original outbred population, which allele becomes fixed (homozygous) in a particular inbred line is partly a matter of chance and partly of natural selection, and it is this genetic variation between different inbred lines which can be used as a tool to analyse behaviour. Because the members of each inbred line have identical genotypes, animals with phenotypes which are to some extent predictable can be produced in large numbers. This is not to say that members of the same inbred line have identical phenotypes; they do not and the individual variation within inbred lines can be very large (Lerner 1954). Nonetheless, if inbred lines differ in the mean value of some behavioural character then this difference can be used as the basis for behavioural analysis.

Selected lines are produced by choosing as parents in each generation those individuals which show the most extreme values of some behavioural character. Selection experiments nearly always work. Lines selected for high and low values of a character gradually diverge over generations, showing that much of the continuous phenotypic variation seen in outbred populations has some genetic basis. Examples of characters successfully selected in this way include spontaneous activity in rats, alcohol preference in mice, mating behaviour in chickens and geotaxis in *Drosophila*. As in inbred lines, there may be individual variation in phenotype within selected lines, but the mean differences between lines selected for different values of a character provide a useful basis for further behavioural analysis.

Another aspect of quantitative genetic variation can be used to study behavioural organisation. If two inbred or selected lines are crossed and the resulting F_1 offspring are then mated together, the various gene loci which differed between the original lines will start to segregate in the next generation, F_2. Thus the F_2 can be used to ask whether characters which differed between the original two strains did so because they were affected by genetic variation at the same loci, or whether instead they were associated by chance in the original strains. For example Batty (1978) demonstrated that, when she compared eight inbred strains of mice, there was a negative correlation between the level of circulating testosterone in males and various measures of male sexual behaviour. To determine whether this association was likely to be due to chance, Batty examined an F_2 generation and again found a negative correlation, suggesting that the two characters are indeed affected by genetic variation at some, at least, of the same loci.

1.4.3 *Audiogenic seizures in mice*

One of the most successful cases of the use of quantitative genetic variation to analyse behaviour has involved the use of inbred and selected lines to study audiogenic seizures in mice. This example is discussed at length by Fuller and Thompson (1978).

Audiogenic seizures are shown by some mice in response to a loud noise. Where a complete seizure occurs, immediately after the loud noise the mouse startles, remains quiet for a short time and then starts to run wildly. The running may then give way to normal behaviour or to a clonic seizure, in which the mouse falls on its side and shows a running type of limb movement. This clonic seizure may give way to a lethal tonic seizure. Audiogenic seizures are not restricted to mice; they have also been reported in rats and rabbits.

Different inbred strains of mice differ in the proportion of individuals which show audiogenic seizures, in the age at which seizures are most likely and in the severity of the seizures shown. This pattern of differences, together with the results of selection experiments, makes it seem unlikely that a simple Mendelian pattern of inheritance of seizures is involved, and indeed most genetic studies where different strains are crossed indicate the involvement of many gene loci in producing the different patterns of seizures shown by different strains.

The neural basis of audiogenic seizures is unknown, but inbred and selected lines have been used to investigate the association between the susceptibility of a strain to audiogenic seizures and the level of amine neurotransmitters in its brain. A comparison of two inbred strains of mice revealed that the strain which did not show seizures (C57BL) had higher levels of brain norepinephrine and serotonin (5-hydroxytryptamine) than mice from a strain that did show seizures (DBA/2) (Fuller & Thompson 1978).

Genetic techniques for studying behaviour are often most useful when combined with the use of other techniques, and this is well illustrated by work on audiogenic seizures. Administration of a drug known to deplete brain amines increased the likelihood of audiogenic seizures in both the above strains, while increasing the level of brain amines by administration of a drug that slows their breakdown had the effect of protecting against seizures. Further studies along these lines have indicated that it is serotonin rather than norepinephrine which is the critical substance affecting seizure susceptibility.

Much work remains to be done on the reasons for variation in the time of onset of seizure susceptibility and in seizure severity.

1.4.4 *The use of quantitative genetics to deduce the history of natural selection on a character*

The genetic variants that we find in natural populations are not a random sample of all those which have occurred; the survivors have passed the test of natural selection. Mather and Jinks (1971) particularly have been responsible for pointing out that the patterns of quantitative genetic variation we find for particular characters may tell us how selection has acted in the past.

The details of the genetic arguments involved are extremely complicated, but the basic idea is a fairly simple one. Suppose we have a character, such as male mating ability in *Drosophila*, which is under directional selection. This means that because the ability of a male fly to obtain matings with females is an important component of his overall fitness, it will be under selection to increase. Any new mutations which increase male mating ability, without correspondingly lowering some other aspect of fitness, will be favoured by selection and will spread to fixation. This means that we would expect to see very little heritable genetic variation for male mating ability in wild fly populations; it should have been exhausted by generations of natural selection. There is in fact evidence that the type of genetic variation, called the genetic architecture, underlying male mating ability does bespeak such a history of directional selection (Fulker 1966). In contrast, other characters, such as the number of bristles on the abdomen in *Drosophila melanogaster*, show high inherited variation, suggesting a history of weak selection.

These arguments need to be treated with slight caution. For example, there may sometimes be strong selection in favour of heterozygous individuals; this will lead to inherited variation for the character in question, which would not here indicate weak selection. In addition, selection for a character like male mating ability may well eventually result in lowered fitness in other respects, for example longevity may be affected. Animal design, including behavioural design, probably involves many trade-offs and compromises, and to consider the effects of selection on a single character in isolation from the consequences for other characters is tempting, but probably dangerous.

1.4.5 Conclusion

Quantitative genetic variation is of particular interest to behavioural scientists because it is the type of variation often found in human and other animal populations. If we want to understand the genetic variation underlying much biological variability we therefore must often use quantitative genetic techniques.

These techniques do have some disadvantages; producing inbred and selected lines can be slow and laborious, and the mathematical techniques involved are often far from simple. The use of F_2 generations from different inbred and selected lines (see section 1.4.2) is a particularly promising approach, because a range of variation can be generated in one cross, whereas the same range of variation could only be produced between inbred or selected lines with far more labour. The other main problem with quantitative characters is that the differences involved can be subtle and difficult to measure. It is perhaps for these reasons that quantitative variation has not been so thoroughly exploited as single-gene differences in analysing the development and functioning of the nervous system. Nonetheless, the number of such studies is increasing, and the interaction of these with studies of genetic architecture should do much to explain the patterns of quantitative genetic variation we see.

1.5 Do genes control behaviour development?

Behaviour genetics is replete with controversies. Some of these undoubtedly stem from poor communications between the very diverse interests involved (Feldman & Lewontin 1975). There are probably two important factors here, in addition to the relatively trivial one that scientists sometimes think that their question is the only important one. First, there are often some quite subtle ideas involved and this leads to both misunderstandings between workers and over-interpretation of data. The nature/nurture debate is one area where both of these problems crop up repeatedly (see Lehrman 1970 and Chapter 2 for a discussion of the concepts involved). Secondly, because some behaviour geneticists work on humans, the scientific debates are greatly intensified and other ideological problems occur.

We shall focus on one particular problem that has led to controversy, namely how we should talk about the role of genes in neural and behavioural development. The role of genes in the development of the nervous system, and hence behaviour, has been given great prominence by some authors, a standpoint described by Stent (1980, 1981) as 'the ideological approach'. In Stent's own words: 'The ideological aspect of the genetic approach confronts us with the basic belief that the structure and function of the nervous system, and hence the behavior of an animal, is specified by the genes which are held to "contain the information for the circuit diagram" (Benzer 1971)'.

Some other quotations clearly illustrate this viewpoint.

> 'The circuit components of behavior, from sensory receptors to central nervous system to effector muscles, are constructed under the direction of the genes' (Hotta & Benzer 1972).
>
> 'How genes might specify the complex structures found in higher organisms is a major unsolved problem in biology' (Brenner 1974).

> 'Work with single gene mutations in a variety of organisms is yielding insights on how genes build nerve cells and specify the neural circuits which underlie behavior' (Quinn & Gould 1979).

Consider briefly what might be meant by such statements. It is quite clear that genetic *differences* between individuals can produce *differences* in their behavioural development. What is much less clear is the extent of the role of genes in the development of the behaviour of an individual.

Genes quite clearly do not contain a blueprint or circuit diagram for the nervous system. Blueprints and circuit diagrams are isomorphic with the structures they represent. As we have seen, the nervous systems and behavioural repertoires of animals cannot be broken into units or modules each of which is represented in a single gene or group of genes. Nor does each gene affect only one 'character'. In Dawkins' (1982) well-chosen words: 'The genome is not in any sense whatsoever a scale model of the body'.

If genes are not a blueprint or scale model, then in what sense could they direct development? Brenner (1974) has suggested that they might embody a program for the construction of the nervous system. This must mean that, in some sense, a sequence of instructions, isomorphic with the sequence of developmental events, is present in the genes. Dawkins (1982) has expressed a rather similar view of development: 'The genome .... is a set of instructions which, if faithfully obeyed in the right order and under the right conditions, will result in a body.' Dawkins then goes on to make an analogy between the genome and a cake recipe.

Although the recipe is a better analogy than the blueprint, because it emphasises that there is no one-to-one mapping between the genome (the recipe) and behaviour (the cake), nevertheless the recipe has shortcomings as a metaphor for the role of genes in development. A recipe is the only source of information for the naive cake-maker, who is otherwise dealing with uninformative and passive ingredients and utensils. Development is not like this; genes are not the only, or possibly even the main, source of information and they are not directing the activities of passive or unchanging ingredients.

In the early development of most animals, the cytoplasm of

the zygote can be shown to be heterogeneous in a way that will determine or bias the future body axes of the organism. Other influences such as the site of entry of the sperm into the ovum can also be important in determining body axes (Gerhardt *et al.* 1981). During development the cells of the embryo acquire different fates, in other words they become committed to give rise to particular structures. Which structures they give rise to depends not only on their lineage but also on their location, through interactions with their neighbours. Such a system of spatial information can be demonstrated by experiments where the fate of groups of cells is altered by moving them to different sites or by the removal of adjacent tissue. It is quite possible that this spatial information lies solely in the effect on one cell's genes of the gene products of other cells, but there is no evidence that this is the case and there are other possibilities. Until we can be sure that the events that occur during development can be accounted for solely in terms of the sequential synthesis of different gene products and their subsequent spatial interactions, it may be a better research strategy to regard genes as one of a potentially large number of important influences on development. Certainly, to suggest that genes alone can build nerve cells is misleading; nerve cells and all other cell types arise only by division of pre-existing cells.

The external environment can act as an important source of information during behavioural development (see Chapter 2 and Lehrman 1970). The interactions between developing organisms and the environment can be very intricate. Adherents to the 'ideological approach' might reply that organisms are programmed by genes to respond in particular ways to particular environmental influences. This argument makes the assumption (and it is an assumption, as explained above) that any influence on development which is not environmental must be genetic. It also raises another more general issue, namely: under what circumstances is it useful to think in terms of complex developmental events as being caused by genes? For example, in the cat visual system, a precise pattern of visual experience is necessary for the development of binocular vision. The two eyes must receive patterned visual input simultaneously during a certain critical period (Blakemore 1977). One could describe this situation by saying that the cells in the cat visual system are genetically programmed to form particular synaptic connections according to the pattern of electrical input they receive from neurons further down the visual pathway. However, control of the response to visual input by genes and solely by genes has not been demonstrated, and anyway this description ignores the very important influence of the current state of the whole developing system on subsequent events in development. Rather than abstracting one component of a chain of events as being *the* causal antecedent of that series of events, it seems likely that it will be more productive to study behavioural and neural development with a variety of possible mechanisms in mind. However we think about the problem, it is not going to be an easy one to solve, but if we think solely in terms of genetic influences we may distort our perceptions of events and fail to think in terms of processes occurring at different levels of organisation.

1.6 Behaviour genetics and natural selection

Variation in behaviour within and between populations of a single species is a common finding. As we have seen (section 1.4), this variation in behaviour often has a genetic component.

Why does such genetic variability occur? This is a question often asked by population geneticists, not only about behaviour but also about genetic variation affecting other characters. One possible answer to the question is that natural selection may be involved and we have already touched on this topic (section 1.4.4). Several different sorts of selection are capable of maintaining genetic variability (Parkin 1979).

1.6.1 Frequency-dependent selection

One such form of selection has long been familiar to population geneticists (Clarke 1979), but has only recently come to prominence in behavioural studies. This is frequency-dependent selection, which occurs when the fitness of a genotype depends upon its frequency relative to other genotypes at the same locus. Negative frequency dependence occurs when the fitness of a genotype declines as its frequency increases, and this type of selection can have the effect of maintaining more than one allele at the same locus in a single population.

Several recent behavioural studies have raised the fascinating possibility that selection acting on behavioural variants may often be frequency dependent because the selection on one behavioural variant will often depend upon what the rest of the population is doing. For example, Cade (1979, 1981) has found that some male field crickets (*Gryllus integer*) call to attract females while other males, known as satellites, sit silently near the callers and attempt to intercept females as they arrive. By careful breeding experiments in the laboratory, Cade was able to show that the behavioural difference between the males has a genetic basis. It is not clear whether the fitnesses of the two sorts of male are equal, as would be expected in a system at equilibrium under frequency-dependent selection. (Strictly speaking it is equal fitnesses of alternative alleles which would be expected; this is difficult to assess because many gene loci are involved in producing the behavioural difference between the two sorts of males.) Cade had evidence that calling attracted parasitic flies as well as females, so that subtle observations of life histories would be necessary to measure fitnesses. Intuitively, it seems likely that frequency dependence is operating, because if few males call then few females will arrive, so that calling will be favoured, while if a lot of males call, the benefits of becoming a satellite may well increase.

One problem with these 'alternative strategies' is that one may not be dealing with a genetic polymorphism at all or, if one is, then the selection may not be frequency dependent. It often happens that young, injured or weak individuals show different behaviour; they may not attempt to breed, for example. This sort of behavioural variation is unlikely to be the result of frequency-dependent selection, and is more likely to be a case of 'making the best of a bad job' (Krebs & Davies 1981). Of course, individuals may be injured or weak for genetic reasons, but the selection acting will not be frequency dependent and will instead tend to eliminate the less fit genotype whatever its frequency.

The whole question of alternative strategies is a very interesting one and will undoubtedly attract a great deal of research effort in the immediate future.

A rather better documented type of selection occurs where selection varies between the different parts of the range occupied by a single species, giving rise to genetic differences between different populations. This type of selection is discussed in the next two sections.

1.6.2 *Spider territoriality*

The desert spider *Agelenopsis aperta* is a web-builder and an individual spider defends an area around its web from other spiders, in the sense that other spiders may not build webs within the defended area. The species is found in a variety of desert habitats and territory size varies between habitats. The largest territories are found in areas where the microclimatic characteristics restrict the time during the day when the spiders can feed, because of very high temperatures by day and extreme lows by night. Territories are also large in those areas where prey abundance is low (Riechert 1978). It is highly likely that territory size is adapted to local feeding conditions, spiders defending large territories in areas where a large catchment area is necessary to provide enough prey for one spider during the period available for feeding.

The match between territory size and feeding opportunity could occur in two ways. First, an individual spider could sample the local feeding opportunities and adjust its territory defence in accordance with its requirements. Alternatively, the response could be unlearned and the differences in territory size between the different populations would then be a result of inherited, probably genetic, differences.

Riechert (1981) has performed experiments which strongly suggest that the second alternative is correct. Spiders were taken from two desert habitats: riparian, where territories are small (0.6 m^2), and grassland, where territories are large (3.8 m^2). The spiders were all then accustomed in the laboratory to a prey density corresponding to the average of those encountered in all habitats in the field. Following this, the spiders were placed individually in small metal cylinders from which they could not detect the presence of the other spiders, and here they set up webs much closer together than in the field. The cylinders were then removed and the spiders' behaviour observed.

The results of the experiment showed that spiders from the grassland population set up their usual large territories while spiders from the riparian habitat set up small territories, strongly suggesting that experience of local feeding conditions does not affect territory size. In addition, the territories set up by spiders from both populations resulted in some individuals being unable to obtain a territory, and the results indicated that high spider density does not result in a decrease in individual territory size. Similar findings were made with spiders reared from the egg in the laboratory. In the wild, the territorial behaviour can result in some individuals failing to hold territories and hence failing to breed. Riechert (1981) has suggested that in both habitats territory size is adjusted to the poorest feeding conditions encountered in that habitat, ensuring an adequate supply of the resources necessary for survival in bad times. Some flexibility is built into the system by the size of the feeding web, which is changed in response to current prey densities.

1.6.3 Feeding behaviour of garter snakes

In coastal California the garter snake (*Thamnophis elegans*) takes more than 90% of its diet as slugs. Outside the range of slugs, in inland California, *T. elegans* feeds mainly on frogs and fish (Arnold 1981).

These food preferences are not learned; most newborn coastal snakes attack slugs on first exposure, while most newborn inland snakes refuse slugs and will starve to death if not given an alternative food. By crossing inland and coastal populations Arnold could demonstrate that the young snakes did not particularly resemble their mother as opposed to their father; these snakes are incubated inside their mothers, leading to the possibility that the mother's diet could directly influence the young, but the lack of maternal effect in the interpopulation crosses ruled out this explanation.

These results strongly suggest that the garter snakes in different geographic areas have been selected to be responsive only to those prey types encountered. The difference between coastal and inland populations is one of degree; both populations contain some individuals which are responsive and some which are not responsive to slugs.

1.6.4 Conclusions

It has only been possible to touch on a very small aspect of the evolutionary applications of behaviour genetics. One of the fascinating features of genetic variation affecting behaviour is that the behaviour may in turn have genetic and hence evolutionary consequences. For example, a genetic variant that influences habitat selection will result in an organism inhabiting a different environment. This will in turn alter the type of natural selection acting on the organism so that other aspects of the phenotype may become modified over generations (Partridge 1978).

Genetic variants affecting mating behaviour can have similarly far reaching consequences. For example, a tendency to mate on a particular food source has been implicated in the process of sympatric speciation (speciation without geographic isolation) in insects (Bush 1969). Changes in mating behaviour are probably also very important in producing reproductive isolation during the process of allopatric (geographical) speciation.

1.7 Selected reading

Because of the great diversity of approaches involved, the literature on behaviour genetics is very widely scattered through books and journals. Two useful and readable general reviews are provided by Manning (1975, 1976).

The genetic approach to the study of the nervous system in invertebrates is reviewed by Macagno (1980). The use of mutants to study the nervous system is critically appraised by Stent (1981). Burnet and Connolly (1981) discuss the role of genetics in the investigation of behavioural organisation. Alternative strategies are considered by Davies (1982).

CHAPTER 2
GENES, ENVIRONMENT AND THE DEVELOPMENT OF BEHAVIOUR

PATRICK BATESON

2.1 Introduction

The study of behavioural development has attracted some of the most bitter and protracted controversies in the whole field of animal behaviour. Some people have argued for the overriding importance of the genes and others have insisted that the environment plays an enormously important role in shaping behaviour. Usually, but by no means always, the genetic determinists were biologists and the environmentalists were psychologists or social scientists. By degrees both sides in the nature/nurture dispute have come to appreciate that the question of what determines the character of a behaviour pattern cannot be expected to produce an either-genes-or-environment answer. Most biologists would agree nowadays that gene expression depends on conditions that are external to the genes and often external to the organism. Most psychologists would accept that environmental influences on behaviour operate through bodies which are themselves greatly affected by their genes. However, all is not sweetness and light yet. A persisting and controversial view is that behaviour can be usefully divided up into genetic and environmental components. Therefore, the first part of the chapter is devoted to the issue of innateness, the notion that some behaviour can be completely coded for in the genes. Much of the disagreement arises from markedly different views about how developmental processes work. When these differences are brought into the open, it is much easier to see how the argument can be resolved.

Resolution can also be obtained on another long-standing confusion. Behavioural adaptations are so exquisite that they demand an answer to the question of what processes might have generated the beautiful fit between the animal and its environment. An adequate solution is not necessarily provided by merely asking whether or not learning was involved in the development of the individual animal. The question has to be posed differently because behaviour that has evolved in response to natural selection during evolution may, nonetheless, involve learning during its development.

With the major controversies out of the way, the second half of the chapter deals with more straightforward matters. Some of the experimental studies of origins are described and the ways in which internal and external influences can interact with each other are then considered. This brings us, finally, to the issue of underlying regularity in developmental processes and what sorts of rules should be looked for.

2.2 The problems of 'innateness'

Is 'innate' a useful term to apply to behaviour? Disagreements arise because the idea of inborn behaviour has a way of slipping through our fingers when we try to grasp it firmly. As is so often the case when clever people fail to agree, more is at stake than mere definition. The term carries with it a distinctive if not always clearly stated view of how development is thought to take place. Once made explicit, some of the predictions from this hypothesis are open to direct test and, therefore, are relevant to a discussion of how genes and the environment influence behaviour. We shall consider some of these expectations but first deal with the matter of meaning. Even the issue of definition raises some interesting general points about behavioural development.

2.2.1 Definition

The term 'innate' most commonly means 'not learnt' and is used (by those who use it) for behaviour that develops without the individual experiencing the stimuli to which it will respond or without practice of the motor patterns that it will perform. In such usage a sharp distinction is made between the experience that has specific influences on behaviour and experience that has general effects. Obviously food and oxygen, along with a great many other external conditions, are required for normal development of all behaviour. The important role of these non-specific factors is admitted but not regarded as decisive in determining whether, let us say, a bird points its beak upwards or downwards while courting a potential mate. Therefore, the definition of 'innate' is emphatically not behaviour that develops without experience in the broad sense. No such behaviour could exist. 'Innate behaviour', according to the most popular definition, develops without the specific experience that could give the behaviour pattern its particular character.

In principle, it should be possible to identify innate behaviour by systematically excluding likely sources of environmental 'information'. The deprivation or 'isolation' experiment, as it is called, can undoubtedly be of service in eliminating possible explanations. Excluding possibilities, however, can never show precisely how the behaviour developed. Even so, the approach has the merit of being positive and directed. Rather than bother about possible unknown sources of variation, the advice to the experimenter seems sensible: if you consider that something in the environment might be important, take it away and see whether the animal can do without it. However, the apparent straightforwardness of this approach is deceptive, and difficulties in interpretation can arise for a number of quite separate reasons.

In practice, it is very hard to draw a sharp distinction between the experience on which the detailed characteristics of the finished behaviour might depend and experience which has more general effects on behaviour (Lehrman 1970; Bateson 1976a). Secondly, it is often difficult to be certain when an animal will generalise the effects of one kind of experience to what superficially looks like a quite different context (Schneirla 1966; Gottlieb 1973). Thirdly, the animal may have different ways of developing a given behaviour pattern and the isolation experiment triggers an alternative (though perhaps more costly) mode of development, bringing the animal's behaviour to the same point as would have been reached if the animal had experienced the environmental conditions from which it was being isolated (Bateson 1981). Finally, an animal that is isolated from relevant experience in its environment may, nevertheless, do things to itself that enable it to perform an adaptive response later on. An example of such self-stimulation comes from a long series of elegant studies by Gilbert Gottlieb. Normally treated Peking ducklings (*Anas platyrhynchos*) are able to respond preferentially to the maternal call of their own species (Gottlieb 1971). However, if they are devocalised in the egg so that they do not make sounds and thereby stimulate themselves, they do not show the same ability to recognise the calls of their own species (Gottlieb 1976b). The devocalised ducklings can behave normally if they are played a recording of the vocalisations made by other ducklings (Gottlieb 1980). These experiments show clearly that feedback from an animal's own activity can play an essential role in normal development.

The difficulties of finding clear criteria by which innateness can be recognised has prompted attempts to redefine it. Lorenz (1965) suggested that the term should be used for behaviour which was adapted to the environment during evolution. The question of how behaviour has been adapted to the environment is discussed in section 2.3, but suffice it to note here that such a change in usage does not reduce the problem of definition. Cassidy (1979) and Jacobs (1981) have argued that when the environment is varied in particular ways, and yet produces no corresponding variability in behaviour, the activities in question should be termed 'innate'. However, this redefinition suffers from the necessity to prove the universal negative, as does the more common usage. A watertight classification requires an impossibly large number of experiments. The problem is that, having satisfied oneself that variation in particular environmental conditions does not influence behaviour, somebody else can come along and show that a different set of conditions influences the behaviour, or that varying the same set of conditions at a different age has a big effect.

All characterisations of 'innateness' ultimately rest on plausibility. The difficulty is that people have very different views about what constitutes evidence adequate to enable useful classifications to be made. When one person is well satisfied that a given pattern of behaviour is 'truly innate', another will feel that alternative explanations have been shut out prematurely. Therefore, as an aid to classification of behaviour, the term is usually more of a hindrance than a help.

2.2.2 *Hidden meanings*

Behind the arguments over whether or not it is useful to divide behaviour up into innate and acquired components lie two highly distinct views of how development takes place. An attempt is made in Fig. 2.1 to represent two extreme positions. The first is most clearly recognised in the writings of Lorenz (1965), even though he had attempted to move away from defining innate behaviour as 'not learnt'. Lorenz imagined that a simple relationship would be found between the starting points of development and the end points. The second view shown in Fig. 2.1 is derived from the thinking of Schneirla (1966) and Lehrman (1970). The current state of the developing animal influences which genes are

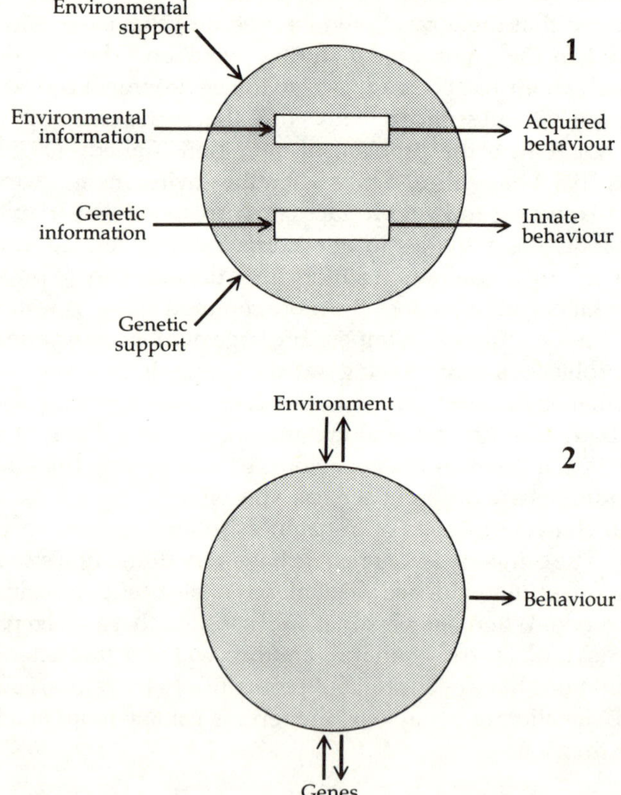

Fig. 2.1. Two views of the ways in which genetic and environmental factors might influence behaviour during development. In studies of animal behaviour, the first view would be associated with Lorenz (1965), and the second view had grown out of the ideas of Schneirla (1966) and Lehrman (1970).

expressed, and also feeds back on to the external world and influences that. The animal is thought to be in a continual state of transaction with its environment, selecting and changing the conditions to which it is exposed.

The advantage of making the different views explicit is that it then becomes much easier to understand what predictions might flow from the two views. Taking the first view, the behaviour that was supposedly dependent on genetically coded information would not be modified by learning once it had been expressed. The accumulating evidence indicates that in its strongest form this prediction is false. It is possible to take a recently hatched laughing gull (*Larus atricilla*) chick and show that it will peck at models of adults' bills. Advocates of the first view would almost certainly want to call the behaviour innate, since the chick had previously been isolated from 'relevant' experience. Nevertheless, as the chick profits from its experience after hatching, the accuracy of its pecking improves, and the kinds of bill-like objects it will peck at are increasingly restricted (Hailman 1967; see also section 3.2).

The first view of development shown in Fig. 2.1 also leads to the expectation that adult behaviour can be analysed into learnt and unlearnt components. Eibl-Eibesfeldt (1970) has argued strongly for the notion of 'instinct–learning intercalation'. Among other examples, he cited his own study of squirrels (*Sciurus vulgaris*) opening nuts, in which a complex sequence can be broken down into components, some of which are learnt and some of which are thought to develop without specific opportunities for practice. This kind of analysis may be profitable sometimes, but in many cases its value seems to be exceedingly dubious (Bateson 1976a; Jacobs 1981). Consider the normal adult songs of three species of American bird shown in Fig. 2.2, and the songs of birds that have been socially isolated at an early age (see also Fig. 3.9). In normal birds which have had their song modified by learning, we can see no obvious component resembling that of the isolated or deafened birds (see Marler 1976). The notion of separable elements of learnt and unlearnt behaviour is actively misleading in these cases, as it probably is in most examples of adult behaviour.

It is not easy to rescue much from the idea that behaviour can be invariably divided up into that which is innate and that which is acquired by learning. However, a more reasonable alternative remains, namely that behaviour varies from examples that are

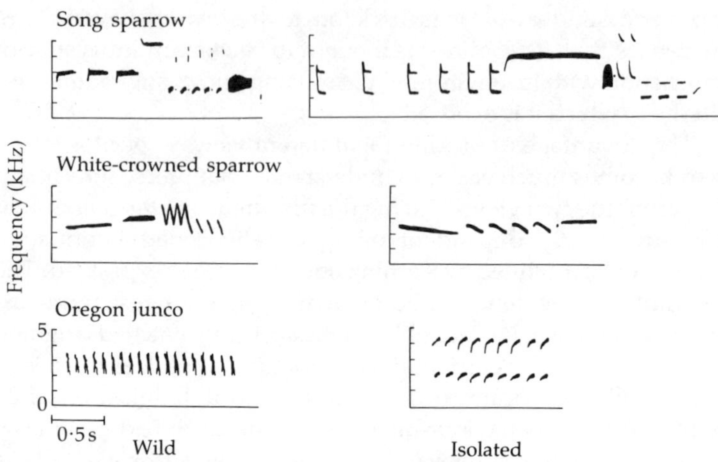

Fig. 2.2. Songs of adult males of three species of birds which have been reared naturally or in social isolation. (From Marler 1976.)

unaffected by learning (though dependent on experience with more general effects) through to examples that are greatly affected by learning. Indeed, Alcock (1979) attempted a four-part classification which ran from Closed Instincts, through Open Instincts and Restricted Learning, to Flexible Learning. This approach may seem sensible, but the problems of definition remain as formidable as ever. For that reason alone such classifications are unlikely to command widespread agreement.

The second view of development shown in Fig. 2.1 does not escape criticism. A model should simplify and point to particular problems that are amenable to analysis. This one primarily seems to tell us that life is complicated. Furthermore, in one respect it can be positively misleading since it implies that the continuous interaction between the animal and its environment modifies behaviour throughout life. As we shall see, many examples of behaviour are greatly influenced by experience at one stage of development but are less affected by similar experience at other stages.

These considerations suggest that we need to move towards conceptions of development that are neither hypercomplex nor so simplified that they grossly mislead. They must take a proper account of the ordered way in which gene expression depends on external conditions, and also allow for the radical transformations in internal state which can arise from experience. In searching for ways or organising our thoughts about behavioural development, it is helpful to employ the metaphor of baking a cake (Bateson 1976a). The flour, the eggs, the butter and all the rest react together to form a product that is different from the sum of the parts. The actions of adding ingredients, preparing the mixture and baking, all contribute to the final effect. Despite the recognisable raisins (if it were that kind of cake), no one could be expected to identify each of the ingredients and each of the actions involved in cooking as separate components in a slice of cake.

2.3 Sources of adaptation

If you watch a bird like a long-tailed tit (*Aegithalos caudatus*) build a nest, its actions seem to be exquisitely adapted to the job of making a strong, cryptic, and well-insulated place in which to lay eggs. Once a site in a bush or tree has been selected, the tit searches for moss and brings it back to the site. When some moss has stuck, it collects spiders' webs and stretches them across the moss. More moss is collected and then more spiders' webs, until a platform has been formed. Now the bird can place moss and webs around itself, building up the sides of the nest. When the nestcup is well formed, the bird fetches lichen and weaves this on to the outside of the nest. Building-up of the sides of the nest is resumed but is periodically interrupted so that more lichen can be added to the outside. Eventually, the bird builds the walls up and over itself to form a dome, but leaves a neat entrance hole in the side. Finally, the nest is lined with a large number of feathers (Tinbergen 1953, cited in Thorpe 1956).

How does the long-tailed tit come to behave in such a well-adapted fashion? For many ethologists, questions about the origins of such adaptations are the most interesting that can be asked about behaviour (e.g. Lorenz 1965). Nonetheless, in as much as it is possible to provide answers, the problem should not be confused with the issue of whether or not learning is involved in development of the behaviour.

In principle, behaviour could be adapted to a particular job, such as that of building a safe, warm nest for offspring, in three separate ways. First, by experimenting on its own with different materials and different actions the bird could assemble the appropriate repertoire for building nests. Secondly, the bird could copy what another, more experienced, bird had done; the process of selecting the actions best adapted to the environment would have gone on in previous generations and been transmitted socially. Thirdly, the birds performing the appropriate actions could have left more surviving offspring than those making inferior nests; consequently, in the course of evolutionary time, genes necessary for the expression of the appropriate actions spread through the long-tailed tit population. In other words, the adaptedness of behaviour arose by the process of natural selection.

It is important to appreciate that all three processes could contribute to the adaptedness of a particular set of behaviour patterns such as nest building. We should not expect to find three classes of behaviour corresponding to the three processes of adaptation. Learning, particularly in complex animals, obviously plays a very important and variegated role in the development of well-adapted behaviour. Among other things, learning helps an individual to predict what is going to happen in its environment, control conditions about it, abandon the performance of behaviour patterns which are no longer serving any useful purpose, and familiarise itself with the details of other individuals and local features of its environment. To tune its behaviour finely to the conditions in which it finds itself generally calls for a considerable capacity of the animal to modify behaviour. Such behavioural plasticity is widespread in the animal kingdom. Similarly, cultural transmission of adapted behaviour is by no means confined to humans (see Galef 1976). One of the first examples to be discovered in other animals was the opening of milk bottles in Britain by great tits (*Parus major*), blue tits (*Parus caeruleus*) and coal tits (*Parus ater*). The spread of the habit from a few scattered locations before 1930 to a great many in 1947 was well documented (Fisher & Hinde 1949; Hinde & Fisher 1950).

Clearly, learning need not be involved in behavioural adaptations that originated in the course of evolution. Hand-reared European garden warblers (*Sylvia borin*) kept in cages all their lives will start to show migratory restlessness at what would normally be the appropriate time in the autumn. They attempt to fly in a southerly direction as is appropriate for a European bird (see Fig. 2.3). The following spring they attempt to fly in a northerly

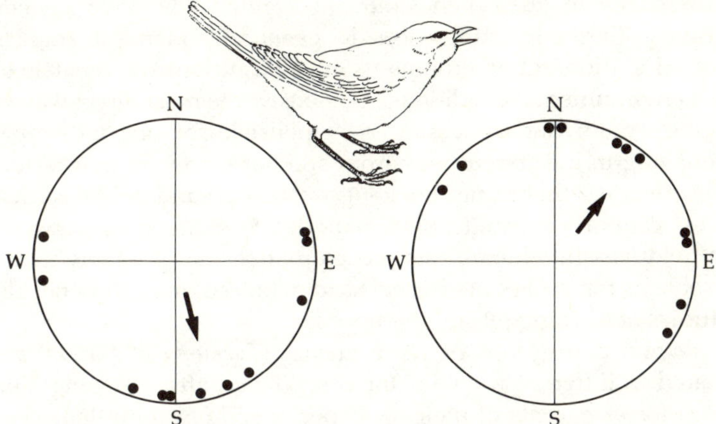

Fig. 2.3. The preferred directions during migratory restlessness of caged hand-reared garden warblers in the autumn and spring. Each dot refers to the mean preferred direction of one bird (Gwinner & Wiltschko 1980).

direction (Gwinner & Wiltschko 1980). Even though the argument rests on plausibility not proof, it is difficult to believe that this behaviour has been learnt. Nevertheless, such remarkable examples do not mean that all cases of behaviour subject to natural selection during evolution develop without specific instruction from the environment. The recognition of kin, for instance, is presumably a form of behaviour that has been selected during the course of evolution. However, by the process of imprinting, kin recognition in many birds and mammals involves a learning process in the course of development (e.g. Immelmann 1972; Bateson 1980). So, even if natural selection is the only source of behavioural adaptedness, it does not follow that no learning is involved in development.

2.4 Origins of behaviour

It will be apparent already that the study of behavioural development is swampy ground. However, one firm plank for research in this area has been to operate on the principle of uncovering sources of *differences* in behaviour, ignoring precisely how those influences operate (Hinde 1968). So, if animals that are known to differ genetically are reared in similar environments, then any differences in their behaviour must ultimately have genetic origins. Similarly, differences in genetically identical animals reared in different environments must be attributed ultimately to the environmental conditions. Obviously, the most direct way to demonstrate that something is responsible for one individual behaving in a different way from another is to vary that factor. Meanwhile, other things are kept constant or randomised so that they cannot contribute systematically to differences between individuals. In general, this procedure does not pose any major problems for studies of external factors. Indeed, many (but not all) studies of learning rely on this approach.

When two groups of zebra finches (*Taeniopygia guttata*) are reared in different ways for the first 30 days after hatching, one with foster parents of their own species and one with Bengalese finch (*Lonchura striata*) foster parents, the sexual preferences of the two groups are dramatically different. Males reared with their own species direct their courtship exclusively towards zebra finch females, and males reared with the other species prefer Bengalese finch females (Immelmann 1972). The difference in the behaviour of the two groups can be attributed unequivocally to the different ways in which they were reared for the first 30 days after hatching. The method of rearing is the source of the variation in the sexual preferences.

By contrast, identification of particular genetic sources of variation is usually less easy than in studies of external sources of difference. Partridge gives examples in Chapter 1 of how single-locus mutants have been used as direct ways of manipulating behaviour. However, such work is in its infancy and recourse to less direct kinds of evidence is almost inevitable.

When related individuals behave more like each other than unrelated individuals, it is easy to conclude that the differences are transmitted genetically from one generation to the next. One way of checking whether or not this is the case is to take the young of one genetic strain and foster them on the parents of another strain. For instance, Ressler (1963) took half the pups of one strain of mouse (*Mus domesticus*) and fostered them on the parents of another strain. When the young grew up they were first given a test in which the frequency with which they pressed a hinged panel was measured. Measurements were then made of the frequency with which the mice pressed the same panel when doing so caused a light to be turned on for one second. The results of this experiment are shown in Table 2.1. While the first measure of panel pressing was unaffected by the identity of the foster parents but was influenced by the genetic strain of the pups, the frequencies of turning on the light were strongly influenced by the foster parents. In the case of the second measure, therefore, transmission from one generation to the next was not genetic. Clearly, similarity between the behaviour of parent and offspring can be explained in a variety of ways.

Table 2.1. The mean frequency per 15 min with which two strains of mice pressed hinged panel when nothing happened (manipulation) or when a light came on (visual exploration). Half the mice had been reared by their own strain and half by the other strain (Ressler 1963).

	Pups	C57BL foster parents	BALB foster parents
Manipulation	C57BL	38	38
	BALB	61	78
Visual exploration	C57BL	62	85
	BALB	73	87

Before leaving the question of genetic influences, it is worth making a brief comment on the shorthand language of population genetics, which is used a great deal in discussions of evolution. In such language one might talk, for instance, about the gene for licking offspring. What is meant by this is that a particular allele makes a difference between an animal licking its offspring and not doing so, if other things are equal. The gene that makes a difference is not sufficient for the expression of licking because it obviously works with a great many other conditions necessary to produce the behaviour pattern. The developmental process generating behaviour must be influenced by many features of the environment and may involve learning. If the environment changes, the nature of the expressed behaviour may be dramatically altered.

The trouble with the gene-for-character language is the implication that a behavioural character spreading through a population in the course of evolution is somehow represented in miniature form in the relevant gene. This encourages a naive form of preformationism and leads people to suppose that developmental processes are somehow trivial and uninteresting. Probably the only way in which to clear up the muddle is to abandon the shorthand and always refer to the 'gene that makes a character distinctively different'.

2.5 Heritability

While it is commonly accepted that environmental and genetic sources of variation can both play a part, it is tempting to ask how much of the variation in a population can be attributed to the genes. An estimate of *heritability* is an attempt to provide an answer. The technique is based on a set of assumptions about the way in which the genotype and the environment produce variation in the phenotype. Figure 2.4 shows how a character in two different genotypes might vary across a range of environments. It is clear that in this simple (and hypothetical) case the total variability in the character can be attributed in part to the difference between the two genotypes, and in part to variability in the environment. (When calculating heritability, the statistical measure of variance is used rather than the total range as shown in Fig. 2.4.) The heritability estimate is the variance in a character due to the variation in the genotypes, divided by the total variance in the character (*G* divided by *V*).

Heritability estimates have clear uses in controlled conditions, but they can also be misleading (see Feldman & Lewontin 1975). If, as is usually the case, only a limited range of environments has been used in obtaining the estimate, then the total possible variance in the character would be underestimated and the heritability estimates would seem larger than would have been the case if the phenotypes were obtained from a wider range of environments. Conversely, if only a limited range of genotypes had been sampled, then the heritability would seem smaller than if a wider range of genotypes had been used. A more serious point is that a stated value for heritability assumes that the effects of the genes and the effects of the environment add together to determine the value of the character. It assumes, in other words, that the two lines shown in Fig. 2.4 are parallel. It is not hard to think of a case in which they might diverge. Imagine an example in which

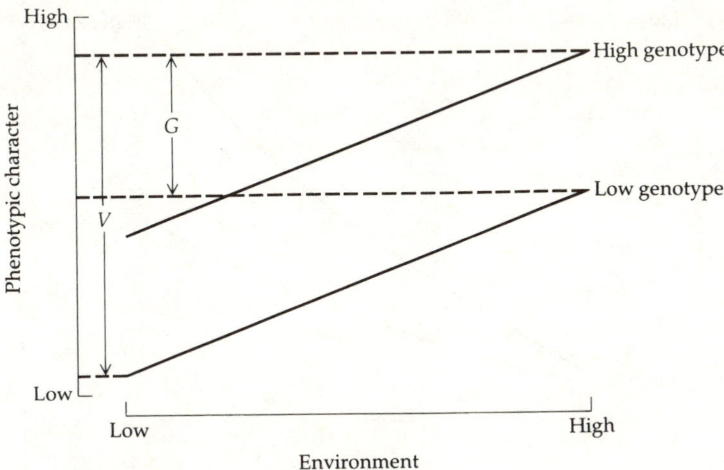

Fig. 2.4. Hypothetical differences in the value of a character for two genotypes when environmental conditions have been varied from those producing low values to those producing high values. If a variety of genotypes had been used, *G* would represent the variance in the character due to variation in the genotype, and *V* would be the total variance in the character.

the character in question is the learning that takes place during the course of exploration. Environments range from those in which there is little food to those in which food is abundant. In environments where little food is available, animals of both genotypes may explore little, but in environments where food is abundant, the variation in exploration may be very great (see Fig. 2.5). In this case, a single measure of heritability would be extremely misleading and would greatly over-simplify the interactions that take place between the internal and external influences on behavioural development.

Even if a heritability estimate could be legitimately used as a rough guide to how much of the behaviour variation is due to genetic variation, it bears no correspondence to the extent to which behaviour has been influenced by learning (see Jacobs 1981). If an unlearnt behaviour pattern had been under strong selection pressure in the course of evolution, all the variation in the genes influencing that behaviour might have been lost. In such a case, any variation in the behaviour would be due to environmental factors and the heritability would be zero. Clearly, the naive

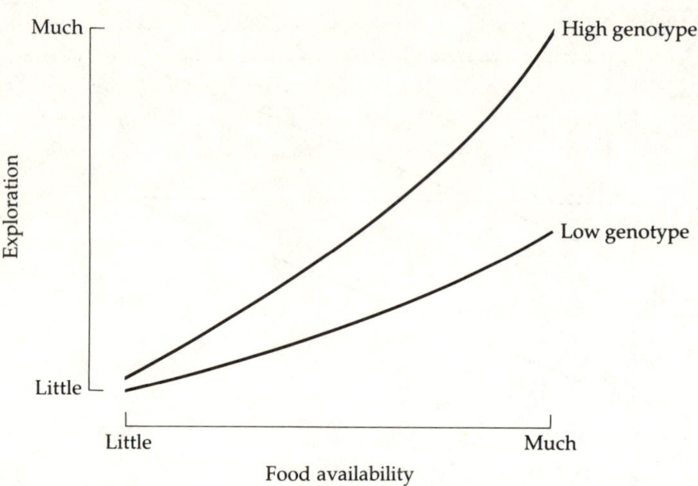

Fig. 2.5. Hypothetical differences in exploration shown by two genotypes when food availability was varied from a little to a lot. The suggestion is that much greater variation is found when food is abundant.

assumption that unlearnt behaviour is necessarily highly heritable would be false, even if we had overcome the difficulties of identifying unlearnt behaviour.

2.6 Indirect pathways

The search for sources of difference in behaviour does not imply that, when the hunt has been successful, an identified source acts directly on the behaviour. A gene may predispose the animal to learn more (or less) and an external event may cause genes to be expressed that would otherwise remain inactive. Some elegant studies of single point mutations in the fruit fly (*Drosophila melanogaster*) have shown that the animal's capacity to establish a link between electric shock and a particular odour is influenced by certain genes. Dudai *et al.* (1976) used the apparatus shown in Fig. 2.6. The flies were put into the lower tube and moved up to the light when it was aligned with a tube above it. The upper tubes had a grid which could be electrified. The training technique involved shocking flies when the upper tube available to them contained one kind of odour and not shocking them when the tube contained another kind. Finally, the flies were tested when neither odour was associated with shock. If they had learned anything, a smaller proportion of flies would go up to the odour previously associated with shock than to the odour which had no such associations. The wild-type flies, on average, did discriminate between the two odours in the tests, even though the difference was relatively small. By contrast, mutant flies appropriately named *dunce* (see section 1.2.4) were indiscriminate. It is by no means certain how the *dunce* mutation affects learning. The mutant flies might not, for instance, have smelt or felt things quite so well and this deficit might have become limiting when they had to do something which even the normal flies found quite difficult. The mutants might have been deficient in some interneurons that were necessary to

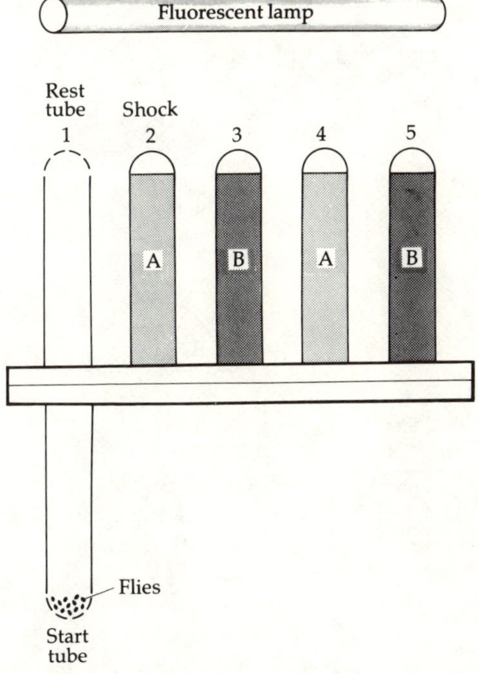

Fig. 2.6. Apparatus used to train fruit flies to avoid shock associated with an odour in tube A. No shock is associated with the odour in tube B. The flies are started from the bottom tube and encouraged to move upwards by placing a light above the upper tube. The flies are exposed, in turn, to the tubes from left to right. When tested with the two odours after training, no shock is used. (From Dudai *et al.* 1976.)

make connections between different sensory modalities, or between the sensory input and the motor output. Finally, the animals might have been defective in arousal or motivation, even though their capacity for storing information was intact. Despite these uncertainties, the fact remains that a single point mutation can influence the ways in which the fruit fly can utilise information from its environment.

Environmental sources of difference can act by triggering the expression of genes which would otherwise have remained repressed in the individual's lifetime. For instance, overcrowding can cause the offspring of non-migratory locusts (*Locusta migratoria*) to grow up to be different from their parents (Dempster

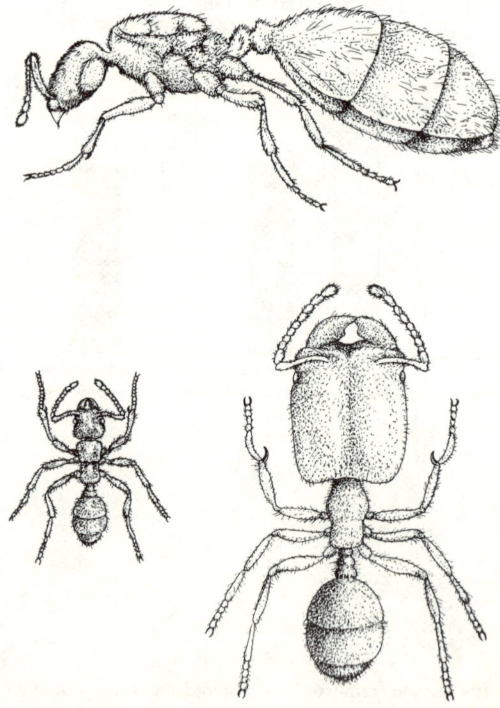

Fig. 2.7. A queen ant (top), a soldier (lower right), and a minor worker (lower left) of the species *Pheidole kingi*. The ants might be sisters but have developed differently because of different treatment when larvae. (From Wheeler 1910, cited in Wilson 1971.)

1963). By degrees, the population becomes migratory. The solitary and migratory forms of locusts are so different that they were once classified as distinct species. Similarly, depending on the nutrition that they received when they were larvae (as well as other environmental factors), female social Hymenoptera can grow into very different adults (see Wilson 1971). Figure 2.7 shows how differently three ants of the species *Pheidole kingi*, which might be sisters, can grow up if they were treated differently when young. Not only do the queen, the soldier and the minor worker look different from each other, but they also behave in quite distinct ways.

2.7 Varieties of influence

Attention is easily riveted on those genes and those experiences that have specific and distinctive effects on behaviour. The alleles that affect the particular aspects of nest cleaning behaviour of honey bees (*Apis mellifera*) are striking because each one influences one aspect of the hygienic behaviour but not others (Rothenbuhler 1967). The early experience that affects mating behaviour of many birds is dramatic because birds with different experiences can later have sexual preferences that are qualitatively distinct from each other (Immelmann 1972). We should not conclude, though, that such influences are the sole stuff from which behavioural clothes are made. Many genes and many forms of experience have non-specific or quantitative effects on behaviour. Most is known about different kinds of experience but the points that follow about the variety of ways in which experience can operate in development also apply, in principle, to genes.

The accuracy with which day-old domestic chicks (*Gallus gallus*) peck at seed can be improved by their being exposed to light for an hour, 3–5 h before they saw the seeds (Vauclair & Bateson 1975). Chicks that were free to move while being exposed to light were most accurate when tested later. However, even those chicks that were prevented from moving their heads during exposure to light were significantly more accurate than the chicks that had never seen the light until they were tested. Clearly, exposure to light influences visually guided behaviour in quantitative ways. However, even in dark-reared animals, the accuracy improves with age (Cruze 1935). It would seem then, that what light does is *facilitate* certain aspects of behavioural development. Eventually, rearing in the dark leads to a regression in the ability of chicks to see (Padilla 1935). So, later on in development, light seems to serve a *maintenance* function.

The influence of light on visually guided behaviour makes another point which has been neglected in previous discussions about the variety of ways in which experience can influence development (see Bateson 1976a, 1978a, Gottlieb 1976a). An experience may not actually trigger a change in behaviour but, nevertheless, it may permit something else to do so. Thus, exposure to patterned light can subsequently allow chicks to learn the difference between red and blue targets (Cherfas 1977, 1978). If the coloured head of a mapping pin has been covered with a bitter-tasting substance called methyl anthranilate, and the chick has pecked once at it, it will subsequently refrain from pecking at the colour associated with the bitterness. Dark-reared chicks, despite receiving the same dose of methyl anthranilate when first trained, subsequently fail to discriminate between red and blue mapping pins. Prior experience with patterned light enables the subsequent training experience to trigger passive avoidance of the target associated with bitterness. Patterned light serves a permissive role without initiating the subsequent change in behaviour.

To summarise, the conditions necessary for the development of a behaviour pattern act in a number of quite different ways; these are shown diagramatically for external experience in Fig. 2.8. The conditions may initiate the development of qualitatively distinct forms of behaviour, they may facilitate processes that are already in operation, they may maintain the end products of development in the behavioural repertoire, or they may predispose the animal to behave in a particular way without acting as a trigger. It must be emphasised that these distinctions apply not only to the external factors influencing behavioural development (e.g. experience) but also to internal ones (e.g. genes). A further point is that all four kinds of external and internal influences on behaviour can in principle range from those that have specific effects to those that have general effects (Bateson 1976a). What we end up with, then, is a large number of different kinds of influence. It seems reasonably likely that many patterns of behaviour in complex, long-lived animals will be affected by a combination of conditions from all the categories of influence.

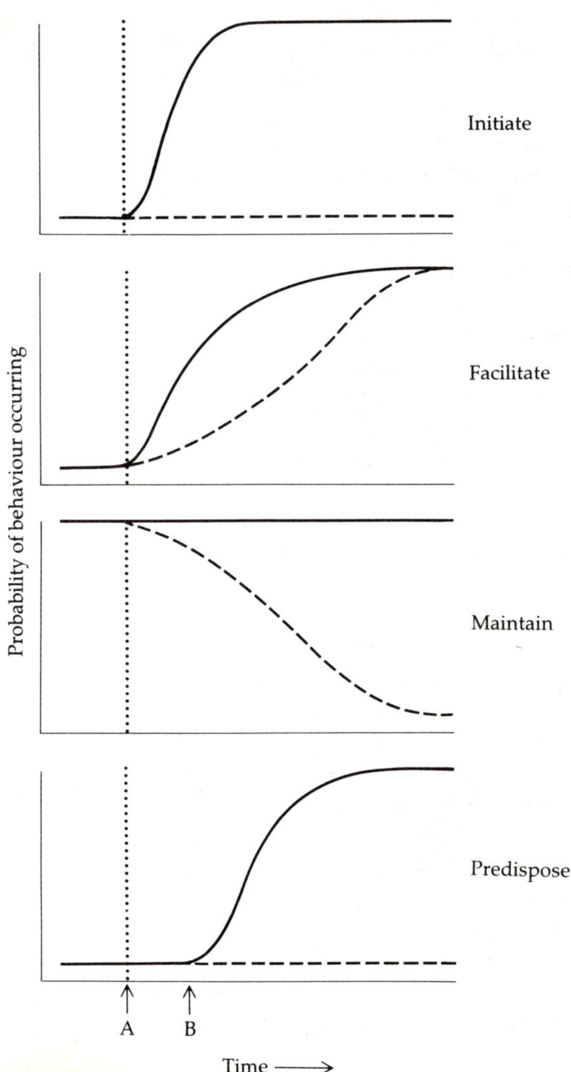

Fig. 2.8. Different ways in which experience with event A might influence the likelihood with which behaviour occurs. In each case the solid line refers to the occasion when event A has occurred, and the dashed line to when it has not occurred. Event B is necessary for the difference in the bottom case (predisposition) but only initiates a difference if it has been preceded by event A. The top three cases are based on Gottlieb 1980.

2.8 Analysis of interactions

Scientists are quite properly taught to look for 'causes'. It is commonly concluded that when an experimental manipulation produces an effect, the manipulation is *the* cause. Other conditions that might also have an effect are ignored or, worse still, regarded as irrelevant. In that state of mind, it is easy to suppose that if a gene has been shown to influence a behaviour, then environmental conditions are unimportant—or vice versa. Even when the narrowness of such conclusions is recognised, many scientists will continue to derive them in the interests of what is misleadingly called 'simplicity'. As a consequence, much time has been spent debating whether nature or nurture is the more important, and a great deal of research has been concerned merely to show that genes can influence behaviour or that the environment can make a big difference.

Consider the case of the novice cook who had been solemnly told by an expert that to make a good cake, the only thing that matters is use of the right ingredients. Feeling that such advice was incomplete, our novice sought a second opinion. The second expert strongly attacked the views of the first and counselled that the baking temperature was all-important. Our novice would rightly feel exasperated by both experts' myopia. The point of this analogy is to emphasise that the cooking processes of behavioural development need to be studied with a real sense of what it means to examine the way that many influences, radically different in character, can work together.

The many internal and external influences on behavioural development work together in ways which generally mean that a given end product could not be readily deduced from knowledge of the whole catalogue of influences. If we are properly to understand development, we have to examine what actually happens. The chapters that follow this one explore different aspects of developmental processes, focusing especially on the ways in which behaviour patterns are changed by external experience. We shall consider here just a single case, the analysis of filial imprinting in birds (see also section 4.2.1).

Most, if not all, complex animals tune at least some of their preferences and habits to external conditions in their own lifetimes. Often these are modified most easily when the animal is young. One of the most intensively studied of the learning processes is filial imprinting in birds. It involves the narrowing of pre-existing preferences to a particular social companion. In the case of filial imprinting in mallard (*Anas platyrhynchos*) ducklings, the narrowing is usually to the duckling's natural mother. In the laboratory, if two equally effective inanimate objects are used, ducklings exposed to object A subsequently prefer it to B when given a choice; similarly, ducklings previously exposed to B reject A in favour of the object with which they are familiar.

The young bird needs to discriminate between the parent that cares for it and other members of its own species because adult females discriminate between their own offspring and other young of the same species and may actually attack young that are not their own. So, if there is danger of encountering other adults, which there is in mallards, the recently hatched duckling must learn to identify its mother as quickly as she learns to identify it. However, completing the learning process at too early a stage in development could be dangerous. Those ducklings that completed imprinting too early would have obtained inappropriate or insufficient information about their mothers. They might not, for instance, have adequate opportunities to explore all facets of their mother and so, having learnt only about the characteristics of, let us say, her back view, would fail to recognise her in side view later on when quick recognition was important. With selection pressures pushing in opposite directions, we should expect the evolution of processes that predispose learning to occur at the optimal stage in development. This expectation, based on functional considerations, is justified by the evidence but the nature of the timing mechanisms is subtle. The image of a clock opening a window on the external world and then shutting it again is not good enough (see Bateson 1979).

Generally, the onset of sensitivity is measured in terms of time after hatching. However, birds can vary by as much as 30 h in the stage of development at which they hatch. In other words, when eggs have been incubated under identical conditions, the time from the beginning of incubation to hatching can vary greatly. It is possible, therefore, to have birds of the same post-hatch age which are at different stages of development, or birds that are at the same stage of development but of different post-hatch ages. The influence on imprinting of the general stage of development can be separated from the influence of experience occurring at hatching and after it. Gottlieb (1961) was the first to use such an approach and a refinement of his experiment was carried out by Landsberg (1976). The results of Landsberg's experiment are shown in Fig. 2.9. Both the age from embryonic development and the age from hatching influence the results. In other words, it does look as though the general state of development of the bird plays a part in the onset of sensitivity, but that the events associated with hatching, or the experiences subsequent to it, also play their part. A part of the increase in sensitivity is attributable to changes in the efficiency of the visual system (Paulson 1965). This being the case, the interaction between internal and external influences is particularly easy to understand. Visual experience with patterned light has, as we have seen, a general facilitating effect on the development of visually guided behaviour. It also serves to strengthen connections in neural pathways (Horn *et al.* 1973). Thus, it is probable that the development of the visual pathways, on which filial imprinting must depend, can be accelerated by early hatching if this means that the bird receives more experience with patterned light than a bird which is still inside the egg.

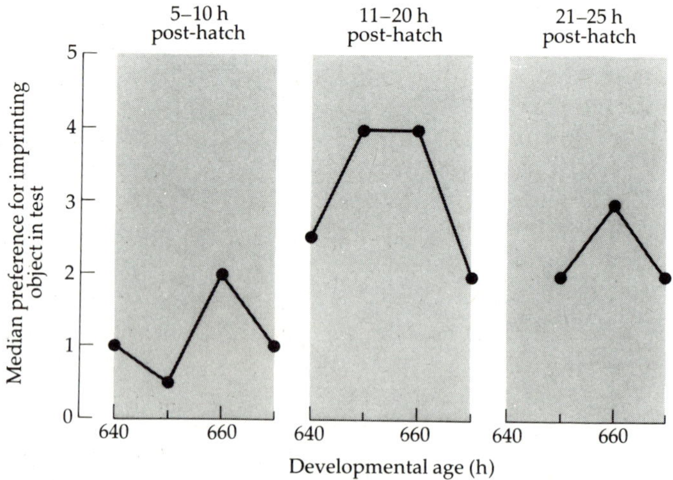

Fig. 2.9. The median preferences of different groups of ducklings for an imprinting object to which they were exposed for 30 minutes at specified developmental and post-hatch ages. Their preferences for the object were measured 24 hours after exposure. (From Landsberg 1976.)

The end of sensitivity to novel objects arises from a property of the imprinting process. When a bird has formed a preference for an object, as a consequence it ignores or even escapes from other objects. Therefore, imprinting with one object prevents further imprinting with other objects from taking place. While some objects are much more effective than others in eliciting social behaviour from naive birds, domestic chicks can form social preferences for suboptimal stimuli such as the static cages in which they were reared, although this takes several days (Bateson 1964). It follows that if some birds are reared with near-optimal stimuli, such as their siblings, and some are reared in isolation, it should be possible to imprint the isolated birds with a novel object at an age when the socially reared birds escape from, or are indifferent to, new things. This has been demonstrated in both domestic chicks and mallard ducklings (e.g. Guiton 1959; Smith & Nott 1970). Here again, we find an interaction between the internal rules of the learning process and the effects of experience. Experience with a particular object coupled with its rule-bound effects on responses to novel objects gives rise to the end of the so-called 'sensitive period' for imprinting. People have difficulties in getting older birds to respond socially to an object because the birds have developed a preference for something else.

While the development of new preferences is initially prevented by escape from novelty or by the low level of social responsiveness to unfamiliar things, enforced contact may wear down these behavioural constraints to the point where the bird does form a new preference. Escape from novel objects can be habituated (e.g. Hinde, Thorpe & Vince 1956; Ratner & Hoffman 1974). What is more important is that domestic chicks that had developed a preference for object A can be induced to prefer object B to object A by a sufficiently long exposure to object B. Salzen and Meyer (1968) found that, when reared for three days with a green cloth ball dangling in their cages, domestic chicks preferred the ball of a familiar colour to a blue cloth ball in a choice test. After a further period of exposure of three days to the blue cloth ball, the birds' preferences were reversed in the choice test. It turns out that their early preference is not forgotten and may resurface after isolation from both types of cloth ball (Cherfas & Scott 1981). So it would seem that protection of preferences from subsequent disruption is accomplished not only by behavioural means but by internal processes as well.

The objects that can be used for imprinting experiments, and the effects they have on the restriction of preferences, also serve to emphasise the interaction between internal and external influences. While the popular notion of imprinting is that attachments can be formed to almost any object, it has long been apparent that the bird's responsiveness is markedly constrained. Some objects are much more effective than others.

This initial bias interacts with the effects of experience. When inexperienced day-old chicks were given a choice between a red and a yellow flashing light, they showed a preference of x units for the red light (Bateson 1978a). After exposure to the red light for 60 minutes, chicks showed a preference of $y+x$ units for red and, strikingly, after exposure to the yellow light for 60 minutes another

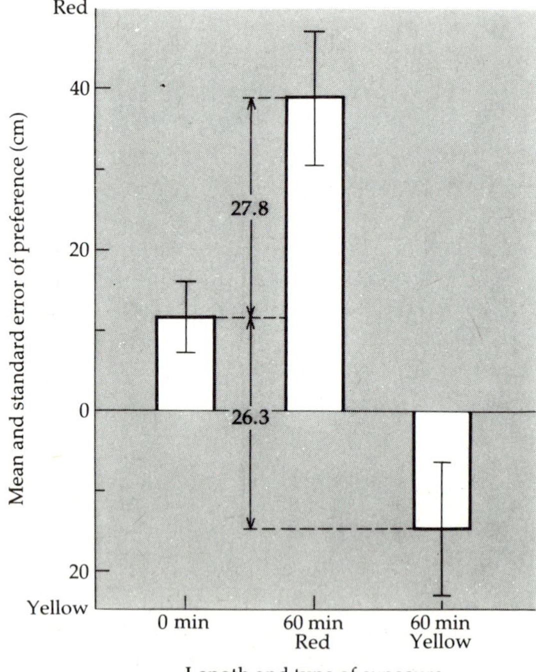

Fig. 2.10. The preferences of day-old domestic chicks when given a choice between a red rotating flashing light and a yellow one. One group of chicks was untrained and the other two were previously trained for 60 min either with the red light or with the yellow light (Bateson 1978a).

group showed a preference for yellow very close to $y-x$ units (Fig. 2.10). In other words, in this experiment the preference of experienced birds was determined by the superimposition of the effects of experience on an initial bias. It is unlikely that the simple additive process suggested by these results will usually operate, and the experience with the initially less effective stimulus would be expected to leave less impact. Nevertheless, the strong hint that preferences are determined by a stable, predetermined bias on the one hand, and by a highly plastic process on the other, is important.

It is clear from these examples of filial imprinting that young birds have strong internal predispositions to, and internal constraints on, what they can learn and when they can learn it. It is also plain that the social preferences of any one bird are strongly influenced by the particular experience it has had. Of course, we cannot directly observe the internal processes at work, and the postulated internal rules controlling what is happening are inferences only. We should not treat them as though they have the same status as external experience which can be directly manipulated and measured.

The example of filial imprinting also illustrates another general point. It is clearly wrong to draw the obfuscating conclusion that in the development of social preferences everything interacts with everything else, or even that the interactive processes determining the character of behaviour operate continually throughout ontogeny. Clearly, the developing animal is more sensitive to some things than to others and its sensitivity also changes as it grows older.

2.9 Regularities

Despite individual differences, members of the same species tend to end up resembling each other in behaviour more than they resemble members of another species. Clearly, behavioural development is very far from being disorderly. It may be complex but it is certainly not chaotic. Something gives development its direction and keeps it in order. Something is responsible for the regularity. There are four separate aspects of this, involving rules for (a) internal integration, (b) compensation for setbacks, (c) external triggering, and (d) external instruction. The likely rules for a particular kind of external instruction (filial imprinting) were considered in the last section. The need for internal integration arises because many perceptual and motor sub-systems arise independently and yet have to be brought together for the efficient performance of behaviour.

Rules for compensation are particularly interesting. If a developing animal is deprived of food or falls ill, it stops growing for a while. If a period of illness or starvation persists, the weight of the young animal may fall considerably behind that of others not suffering such misfortune (McCance 1962). However, the weight of the animal can quickly recover to where it should have been if it is put back on a normal diet or if it recovers from illness. Some kind of internal regulation is clearly implied (see Bateson 1976b). As yet, very little is known about the actual nature of the control mechanism. In some cases it may reside within the parent. If the parent detects that the behaviour of its offspring is different from what it should be, the parent may behave in such a way as to make good the discrepancy. For instance, a mother goat (*Capra hirens*) with two offspring will, unless one of them is very weak, prevent the more vigorous one from suckling first (Klopfer & Klopfer 1977). Presumably if she did not do this, the more vigorous of the kids would get most of the milk and would grow at the expense of the other. This would reduce the reproductive success of the mother. The extent to which the mother compensates for slight deficiencies in her young is particularly obvious if the stronger of the two kids is stupefied with an anaesthetising drug. Under natural conditions her behaviour would act to keep the offspring on a normal developmental pathway (Fig. 2.11).

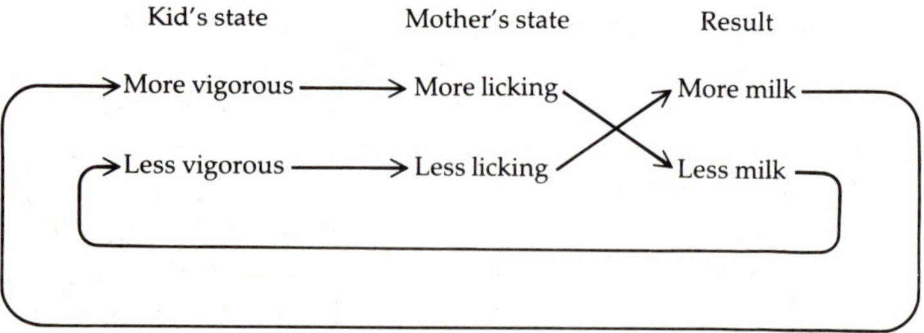

Fig. 2.11. The way in which a mother goat ensures that both her kids grow at the same rate. Licking by the mother makes it more difficult for the licked kid to get milk (Klopfer & Klopfer 1977).

Compensating for shortcomings may have diminishing returns. In the case of the goat, the weaker kid will be neglected by the mother if it has been stupefied, suggesting that there may come a point when she settles for raising one offspring (Klopfer & Klopfer 1977). Similarly, if animals are deprived of food at stages when they are growing very rapidly, they may be permanently stunted. One functional explanation for stunting is that the animal does not endlessly attempt to reach a state which may never be achievable in the particular conditions in which it is developing. Deprivation of optimal conditions for one system does not necessarily imply that conditions are bad all round. Normal development of the animal's other systems may still be possible. Although the animal may be handicapped, its chances of surviving and leaving offspring may not be reduced to zero. The long-term effects of certain environments may not merely mean that the animal is making the best of what it has got. Some of the effects of undernourishment on metabolic rate and foraging strategies could, for instance, reflect specific adaptations to environments with little food in them (Levitsky 1979). This relates to the point about rules for external triggering.

Biologists are becoming increasingly alert to the alternative ways in which an individual may develop (Davies 1982; Dunbar 1982). We have already seen the case of the locust which becomes solitary or migratory depending on whether or not it was crowded when it was young. In this case it seems unlikely that the migratory behaviour is in any sense learned in response to the crowded conditions. It is more as though the animal is like a jukebox in which a particular record can be played when a specific button has been pressed from outside. The finger that presses the button does not carry the information that is on the record. In this sense, the environment triggers an appropriate form of behaviour rather than providing the instruction. Clearly, such external triggering requires internal organisation if it is to operate in an adaptive way.

While the regularity of development implies a considerable degree of internal control and organisation, it should not be supposed that the internal rules are the direct expression of genes. The rules themselves have to develop and, clearly, they represent the workings of an already functional nervous system and body. The extent to which their development involves various kinds of experience raises an entirely separate issue. It is not an easy problem to analyse because, as pointed out in the last section, the rules are not directly observed; they are inferred from what happens to behaviour. Secondly, if, as is quite possible, the problem turns out to be concerned purely with the influence on neural development, the neural changes will have to be related to the behaviour of the whole animal. The practical and conceptual difficulties of doing this must not be underestimated.

2.10 Conclusions

The development of an animal's behaviour is greatly influenced by its genes and its environment. The interplay between the animal and its external world is such that the identification of genetic and environmental components in behaviour will rarely be possible. Furthermore, the practical problems of excluding environmental influences that might have a specific effect on behaviour are very great, so any attempt to label a given pattern of behaviour as 'not learnt' rests on plausibility and may not command widespread respect. Nonetheless, it is clearly possible to unravel some of the complexities of ontogeny, first by identifying the major sources of variation and secondly by directly examining the processes of development. To return to a metaphor which I have used at several points during this chapter, looking at the cooking is an important part of such study. By degrees it is becoming possible to understand the rules that generate the order behind all the surface complexity. The subject still has a long way to go but the prospects for uncovering general principles look more and more promising.

2.11 Selected reading

The subject of behavioural development is itself developing fast and it is not easy to provide access points to the literature which will not date rapidly and which are not too restrictive. There is no solution to the problem of dating, except to take note of the most active authors and, by such means as the *Science Citation Index*, to obtain references to their most recent papers and to those who cite them.

Restrictiveness can be overcome by consulting books with chapters by many authors with different views. Fortunately, two stimulating collections on the topic of behavioural development have been published recently (Immelmann *et al.* 1981; Bateson & Klopfer 1982).

The nature/nurture issue has a long history and Oppenheim (1982) provides a valuable historical account. Those who feel that the learning/instinct dichotomy still has some mileage in it are advised to read Lehrman's (1970) classic paper. This was written in the heat of the battle, and a milder but altogether excellent review of the issues is provided by Oyama (1982). She has also written a penetrating essay on sensitive periods in development (Oyama 1979) which should be read in conjunction with my own review (Bateson 1979). Finally, a lucid introductory overview of the principles of behavioural development can be found in Chapter 3 of Hinde, 1982.

CHAPTER 3
THE DEVELOPMENT OF INDIVIDUAL BEHAVIOUR

P.J.B. SLATER

3.1 Introduction

The last chapter has provided ample evidence that development is a continuing interaction between the organism and its environment. It is a process that cannot be usefully summarised by referring to some aspects as innate and others as learnt; the ways in which genes and environment interact are far too complicated for that. Nevertheless, strategies of development do differ. Some behaviour patterns, which are often referred to as having fixed or stable ontogeny, develop very similarly in a wide variety of natural environments. In a few cases, where experiments have been carried out, such behaviour patterns have also been found to develop near-normally in greatly impoverished laboratory conditions. At the other end of the spectrum are behaviour patterns of flexible or labile ontogeny which are greatly affected by natural or artificial changes in the environment. It is the main aim of this chapter to describe some of the circumstances under which such striking differences in ontogeny have been found and to examine the sorts of influences known to affect the way in which individual behaviour develops. This will be done by describing some well-researched examples. However, it is worth making some general points before getting down to dealing with these case histories.

One very important point concerns the origins of variations in behaviour. It is often assumed that behaviour patterns which vary a lot from one individual to another are those which are less dependent on genetic influences and more on influences from the environment. Similarly, the behaviour patterns most constant throughout a species are sometimes suggested as being those least influenced by the environment: for instance, fixed action patterns, of which the German name, literally translated, means 'inherited coordination'. Such arguments are not well founded. It is equally plausible to suggest that individuals differ one from another because a high degree of genetic variance is either encouraged or tolerated, and that where individuals are similar this is because all share certain aspects of their experience. Only by careful experiment can we discover how animals develop to behave in the ways that they do. As we shall see, the results of such experiments often run counter to the intuitions one might have on the basis of observation alone.

A second point to make concerns natural selection. This is simply to stress that selection acts on the outcome rather than directly on the mechanisms by which that outcome is reached. Learning by experience may be an adaptive way of gearing the behaviour of an animal to its own particular environment, but it is not necessarily the best way. It may not be open to animals with short lives or small nervous systems. Where the first encounter with a predator may be a lethal one, it is also essential that the young animal behaves appropriately without experience of that situation. As we shall see in this chapter, one of the interesting things about development is the discovery of just where specific aspects of experience have an influence and where they do not.

A further important point is that development is a continuing process throughout which the animal is being tested by selection. At every stage the individual must behave in an adaptive manner if it is to survive and, for the young animal, this may mean behaving quite differently from the adult. This is most obviously true in species which undergo metamorphosis and in which changes in behaviour are dramatic. For example, frog tadpoles prefer green to blue, whereas in the adult this preference is reversed (Muntz 1963), probably because of the strong differences in feeding and escape behaviour between adult and juvenile. But the need for behaviour to be adaptive at all stages is also true of other species where metamorphosis is not involved and changes are less abrupt. Even in these, the young animal is not just a small adult, nor is its behaviour simply a preparation for adulthood, although this is certainly one aspect of it (Oppenheim 1981).

Development is, then, a subtle process with many different interacting influences determining the way in which an animal behaves at each particular stage. Such influences may be external stimuli or internal events, such as the secretion of a hormone or the attainment of a particular stage of maturity; they may have rather generalised effects on much of the behaviour of the animal or very specific ones on a single, small aspect (Bateson 1976a). Influences important at one stage may not be so at another: the phenomenon of sensitive periods, widespread in development, is an expression of this. Yet, while each stage in development is built up on the last, there is also a sense in which it is self-regulating and thus ensures an appropriate end-point even if a disturbance blows it off course (Bateson 1976b). The sensitive period for imprinting is prolonged if an appropriate attachment object is not encountered at the right time (see Chapter 4); growing rats, if deprived of food for a few days, will subsequently eat more so that their weight rises to rejoin the growth curve they were following before (McCance 1962).

In this chapter we will consider some case histories to illustrate a few of the complexities of development within the individual animal. The first, that of the development of the visual system and how this influences behaviour, shows that the environment may affect even sensory systems and how they function. In the following section, we will consider work on predator and prey recognition in a number of species to show the variety of influences which may be involved in leading to an adaptive outcome. We will then consider bird song development, one of the classic case histories in this area, which illustrates beautifully the interacting roles of genes and environment in development. Many bird species show cultural transmission of song, young birds learning details from adults that they hear. Cultural transmission is one particular aspect of experience which is especially potent in adapting the behaviour of animals to an unpredictable world, enabling them to learn from others who have already experienced its vagaries. For this reason we shall consider it further at the end of the chapter.

3.2 Variations in development: the visual system and behaviour

The commonest way of studying the development of individual behaviour is by allowing young animals to develop in different environments to see what influence this has on their later behaviour. Experiments of this sort have produced dramatic examples of behaviour which appears normally despite great environmental changes and, at the other extreme, of features which are altered by rather small manipulations of the environment. In the latter category, the visual system provides some especially remarkable examples because it has been possible to identify changes in the nervous system induced by specific external stimuli. Although this is a physiological rather than a behavioural example, such changes are likely to have important consequences for behaviour. The original experiments in this field, by Hirsch and Spinelli (1970), involved rearing kittens wearing goggles such that one eye saw vertical stripes and the other horizontal. The cortical cells fed by each eye were later found to be sensitive only to the orientation that eye had seen. At around the same time, Blakemore and Cooper (1970) showed that young kittens, reared in the dark but placed for a few hours each day in a drum lined with either vertical or horizontal stripes, subsequently had no units in the visual cortex responding to stripes of the opposite orientation (Fig. 3.1). Normal kittens also develop with cells in the visual cortex sensitive to stripes of a preferred orientation, but they have roughly equal numbers of cells for each possible angle. In contrast to this, the preferences of the cells in the experimental kittens had clearly been biassed by the experience they had undergone.

Although the experiments of Hirsch and Spinelli have been replicated, not all efforts to repeat the procedure used by Blakemore and Cooper have succeeded (Stryker & Sherk 1975). This may be because cats wearing goggles cannot see stripes of an inappropriate orientation, whereas animals in cages such as those used by Blakemore and Cooper can do so simply by lying on their sides (Mitchell 1978). Thus, differences in behaviour between kittens might lead them to yield quite different results. Despite problems like this, there can now be no doubt, from the many experiments carried out in the past ten years using several different types of visual deprivation, that experience can alter the visual system of young kittens. But exactly how this alteration takes place is still a matter of controversy (Barlow 1975; Mitchell 1978; Rothblat & Schwartz 1978). At least some cortical neurons sensitive to particular orientations do appear to be present at birth (Hubel & Wiesel 1963). It is not known whether further neurons are non-specific at this stage and adopt specificities later according to their experience, or whether the specificity found in these experiments results simply because cells which are

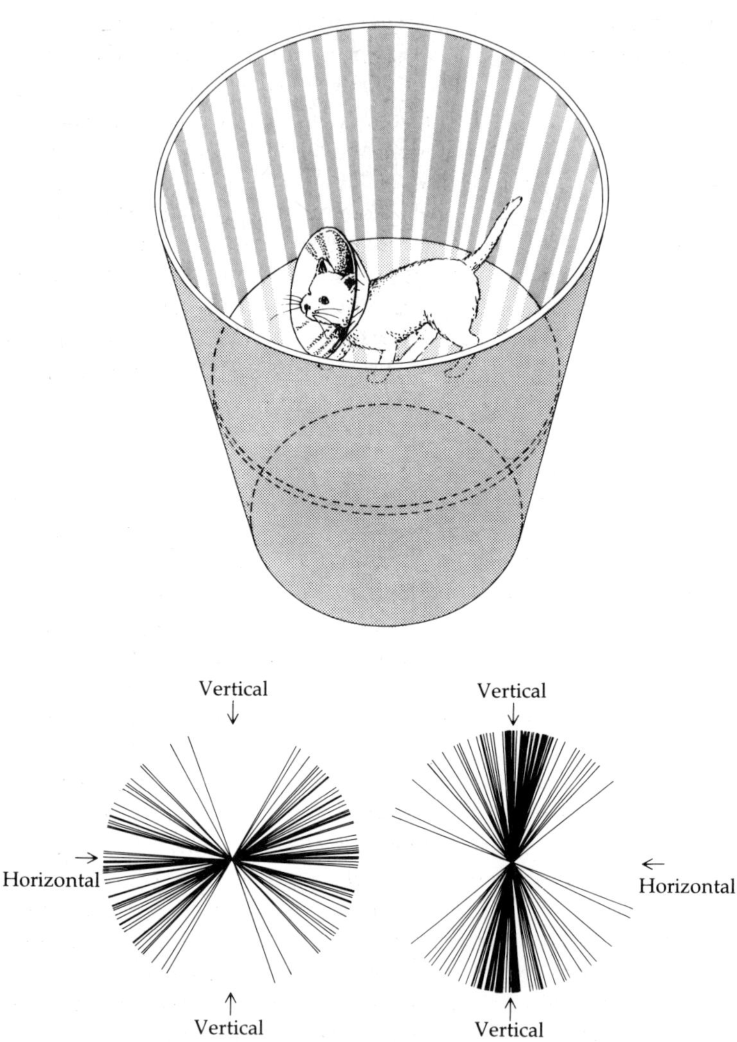

Fig. 3.1. Above: The apparatus in which the kittens studied by Blakemore and Cooper (1970) obtained all their visual experience. In an alternative version, the stripes on the cylinder were horizontal. The collar prevents the kitten from seeing its own body. Below: Diagrams illustrating the orientations of stripes to which units in the visual cortex of these kittens were sensitive. That on the left is for a kitten exposed to horizontal stripes, that on the right for one reared with vertical stripes.

not stimulated with their preferred orientation atrophy during development. Selective cell death could account for the results without individual cells changing their specificity at all.

Why should the cat visual system be so sensitive to environmental influences, especially given that no equivalent effects have been found in monkeys (Wiesel & Hubel 1974)? A clue may be provided by the interesting experiments by Keating (1974) and his colleagues on toads (*Xenopus laevis*). In these animals there is complete decussation at the optic chiasma (see section 1.2.3), the nerves from one eye projecting entirely to the optic tectum on the opposite side of the brain. Stimulating the eye with spots of light and recording from the contralateral tectum shows that the latter contains cells which are sensitive to particular points in the visual field and are laid out over the surface of the tectum on exactly the same plan as the retina. Cross-connections between the two tecta link pairs of cells sensitive to those points on the two retinae which are stimulated by the same object in space. Thus each cell receives a direct connection from the eye on the other side of the head and an indirect one from the eye on the same side. This visual system has made a very important contribution to our understanding of how nervous systems develop. If, at an early stage, the eye of a *Xenopus* tadpole is rotated through 180°, the neurons from it grow back as normal to the correct tectum, and they sort themselves out as they go, so that they link with the same cells as they would have during normal development. The striking finding described by Keating (1974) concerns what happens to connections between the two tecta when this is done to one eye but not the other (see Fig. 3.2). If normal connections were made, one would expect a tectal cell to be stimulated by a particular point on the contralateral retina and, because of the eye rotation, by the point diametrically opposite it on the retina of the other eye. Instead, tectal cells are found to be stimulated by points which are now in equivalent positions in the two eyes. These are the points on the retinae which see the same position in space, so it appears that connections between the tecta are formed in such a way that cells receiving the same input become linked to each other.

Just how such connections are made is a challenging question but, for our purposes, the most interesting point is why the system works in this way. The answer is probably simply that sharp binocular vision is impossible to program into a developing system without such an environmental influence. As *Xenopus* grows from tadpole to toad, the eyes migrate from the sides of the head to the top and front, and their final distance apart depends on how well nourished, and therefore how large, the adult toad becomes. If their final position could be predicted precisely, then visual maps on the tectum which were perfectly in register for the two eyes could perhaps be constructed without using any environmental information. But this cannot be done without the exact position of the eyes being known. While the problem may be an especially acute one for toads, which vary considerably in size, the fine tuning of the visual system is so exquisite in many animals, such as cats and ourselves, that the need for it to depend to some extent

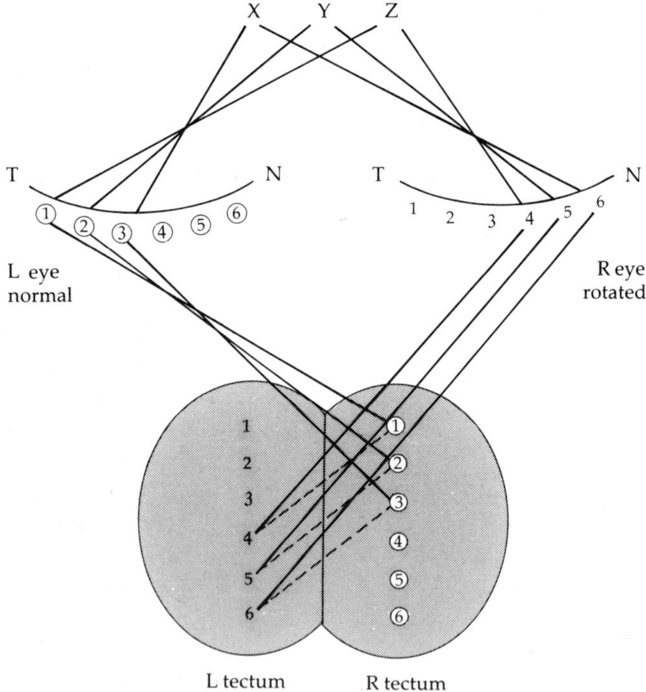

Figure 3.2. Diagram of retinotectal connections, and those between the two tecta, formed in *Xenopus* after 180° rotation of the right eye at stage 32 of larval life. In this eye the temporal (T) and nasal (N) parts of the retina are the wrong way round. Retinotectal connections are formed as if no rotation had taken place but the connections between the tecta link units receiving visual input from the same point in space so that the connections here are different following rotation from those in normal animals. (After Keating 1974.)

on environmental input should not perhaps appear surprising. Furthermore, in growing animals which depend on vision throughout development, the system may have to be retuned and calibrated continuously as the eyes move further apart. Cats may well have a system equivalent to that of the toad, for Schlaer (1971) has shown that the directional sensitivity of binocular neurons in the cortex can be altered by changing the visual input to kittens by using prisms.

Fig. 3.3. A duckling peering over a visual cliff such as is used to test the reaction of animals to heights. The drop on the left is more apparent than real as the whole apparatus is covered with a transparent sheet.

These examples concern plasticity where it might have seemed unexpected, in the development of sensory systems. In contrast, other, no less dramatic, examples can be cited in which the behaviour of young animals appears fully formed even without appropriate experience. For example, young lambs and many other ungulates can walk, albeit somewhat shakily, within 30 minutes or so of birth. Admittedly, they have twice the number of legs that humans do on which to balance, but that hardly seems to account for the fact that we do not usually achieve the equivalent feat until we are over a year old. The contrast is simply in the requirement: human infants are carried by their mothers, while lambs must follow theirs from pasture to pasture. Another example is that of different species of duckling, hand-reared from the egg in identical conditions; those of tree-nesting species are not afraid of heights and leap from them as they normally do when fledging, while those of ground nesters avoid such edges (Fig. 3.3; Kear 1967). These are the sorts of capabilities that ethologists would, in the past, have referred to as 'innate', because they appear the first time the young animal encounters a particular situation. But such a finding does not mean that earlier experience did not affect their appearance or that the behaviour cannot be modified subsequently, both of which are connotations often associated with the word 'innate'. Ethologists therefore tend to

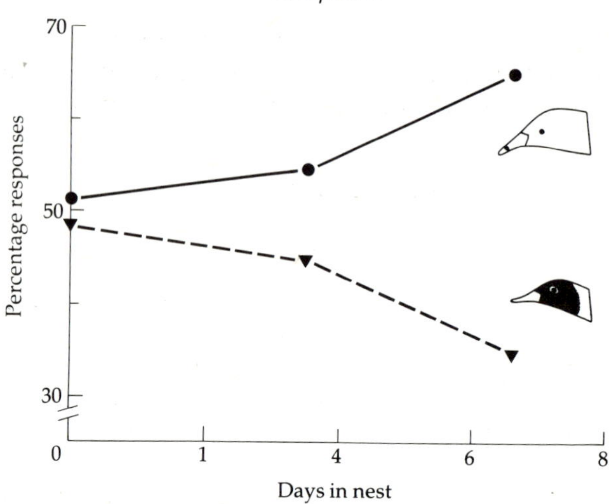

Fig. 3.4. The development of the preference for pecking at a model of their own species (—●—), as opposed to one of the laughing gull (—▼—), in young herring gulls. At birth they peck equally at both models, but after 6 days in the nest they prefer that of their own species. (After Hailman 1967.)

avoid using it nowadays. As well as there being sound theoretical reasons for this (see Chapter 2), there have also been some detailed studies which show just how complex are the interrelationships which lead to fully formed behaviour, even when this appears early in life. A good example of this is the pecking response of young chicks in the herring gull (*Larus argentatus*). On the basis of experiments with nestlings, Tinbergen and Perdeck (1950) concluded that these birds were innately sensitive to exact details of the parent's head and beak. But later studies by Hailman (1967) showed clearly that this sensitivity to exact details appeared only after they had been in the nest for some days, so that they had had some experience of being fed by their parents (Fig. 3.4). Inexperienced herring gull and laughing gull (*L. atricilla*) chicks pecked at models of the other species as much as at their own, but older ones became much more fussy, a process Hailman referred to as 'perceptual sharpening'. However, as with so much else in development, experience is only part of the story. Naive laughing gull chicks do have a tendency to peck at beak-like rods, especially if these are red (as is the parent's beak), if they project down from above and if they move from side to side. These may simply be the stimuli that most strongly excite the chicks' visual system (after all, moving stripes are what stimulate kittens most), but they are sufficient to channel the interest of the young chick so that it pecks at the appropriate stimulus and learns its finer details.

3.3 Avoiding predators and finding prey

3.3.1 Predator recognition

As we mentioned earlier, there are certain stimuli to which animals must respond appropriately the first time that they encounter them, and predators are perhaps the most obvious of these. It may not, therefore, appear surprising that Lorenz and Tinbergen, in experiments described by Lorenz (1939), found that young turkeys (*Meleagris gallopavo*) reacted with alarm when the silhouette of a hawk was 'flown' over their pen. But predators take many shapes and forms: how did the young turkeys know that this particular stimulus was dangerous? Lorenz and Tinbergen argued that it was the combination of short neck and long tail, a configuration shared by most avian predators. In their most dramatic test they devised a model which when flown over in one direction looked like a goose and in the other resembled a hawk (Fig. 3.5); the young turkeys showed alarm only in the latter case.

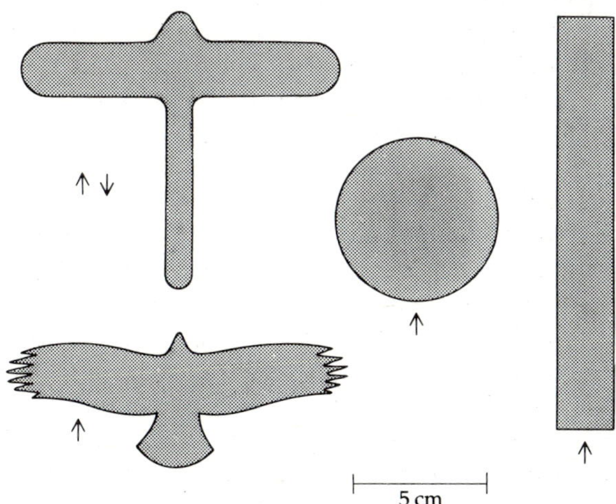

Fig. 3.5. The models used by Schleidt (1961) to test the reaction of young turkeys to overhead objects. Arrows show the direction of motion. The hawk/goose model at top left was sometimes flown in one direction, sometimes in the other.

These experiments have aroused a tremendous amount of subsequent work and a good deal of controversy over the question of whether young birds could respond to such subtle distinctions. Some amazing shapes have been sent flying over pens of fledglings! For example, Schneirla (1965) made the ingenious suggestion that the hawk model was more alarming because it had a broad leading edge and thus made a much more sudden entry on to the retina. If so, he suggested, a triangle moved overhead with one of its edges at the front should induce more alarm than one with a point leading. This was duly tried by Green *et al.* (1968) but no difference was detected. Amongst the most extensive series of experiments was that conducted by Schleidt (1961). He used several shapes, including the hawk/goose model, circles and oblongs (Fig. 3.5), and flew them over naive young turkeys in a long series of tests. He concluded that the birds showed the strongest alarm to the stimuli that they had seen least often. If the same model was used many times, their response diminished by the process of habituation, only to reappear when another stimulus was tried. Appearing suddenly late on in the series, the circle led to more alarm than any of the other models, even the realistic silhouette of a hawk. Schleidt argued that the phenomenon of habituation was sufficient to account for the results that Lorenz and Tinbergen had obtained. Their turkey chicks were kept outside by a lake where they would have seen many geese but few hawks before the tests began.

The idea that young birds respond with alarm to flying shapes that they have not seen before but that they learn, through habituation, to ignore those that occur frequently, is an attractive one which could also apply to ground predators. Given that predators can be lethal, the young animal must respond at once but, as predators take many different forms, it would not be easy to identify all of them reliably at first sight. Instead, the system seems geared to playing safe: assuming the object is a predator until it proves otherwise. However, this is not all there is to it. Some of the most recent experiments using the hawk/goose model have examined the change in heart rate of naive young birds exposed to it, a measure much more sensitive to small differences in alarm than the call rates or fleeing scores most often used previously. Such tests do suggest that there is a genuine difference at this subtle level in reactions to the movement of the model in the two directions (Mueller & Parker 1980). There is thus some evidence that young birds have a predisposition to respond to hawks with alarm, although the effect is not as strong as was originally thought.

The most thorough experiments carried out on predator recognition to date are those by Curio (1975) on the pied flycatcher (*Ficedula hypoleuca*). This species responds to both red-backed shrikes (*Lanius collurio*) and pigmy owls (*Glaucidium passerinum*) with calls and movements indicating alarm (the mobbing response), and these two species also prey upon it. By using models, the characteristics of which were varied, Curio showed that the response of flycatchers involved separate recognition of shrikes and of owls. Generally speaking, the more features of the predator incorporated in the model the greater the response, but characteristics of the two predator species could not be mixed together in a single model to give a stronger effect. Hand-raised birds neglected live members of two non-predatory species, but showed some response to live owls and shrikes though not to models of them. Birds from a population in Spain where shrikes are absent mobbed shrike models only very weakly. The response does not therefore seem to be simply to novelty, but at the same time experience of the predator is not essential for it to be shown.

Attempts to condition mobbing in the pied flycatcher were unsuccessful, and Curio (1975) concluded that experience has little part to play in the recognition of predators by this species. The recognition filters he identified are sufficiently broad to allow response to several species of owls and shrikes. However, in later experiments with blackbirds (*Turdus merula*), Veith, Curio and Ernst (1980) found that this species could be conditioned to mob by the association of harmless objects with mobbing calls. If the object was a bird, the effect was stronger than if it was a plastic bottle, suggesting some constraints, but only minor ones. Just why birds of many species do mob predators is not known, although many suggestions have been made (see Curio 1978). These blackbird findings suggest that one consequence of mobbing may be the cultural transmission of predator recognition. While this may not be a benefit to the mobber, it could be an important means whereby young individuals learn the characteristics of predators.

The evidence therefore points to several different ways in which birds may come to recognise predators. They may be predisposed to respond to certain characteristics without the need to experience them; they may respond with alarm to many novel objects and then, through habituation, learn that certain of them are harmless; finally, they may learn through the alarm of other individuals in the presence of an object that it is dangerous. All these have it in common that the young animal can come to respond to predators without being endangered by them. The fact that learning is involved in the last two cases means also that the individual may be able to respond to the particular spectrum of predators found in its area even if these differ markedly from one place to another.

3.3.2 *Food recognition and handling*

Like predators, potential foodstuffs may come in a great variety of forms, especially for omnivorous species. Where something that might be eaten is actually extremely poisonous, correct recognition without prior experience of it is very important. As with some of the predator examples discussed above, the fear of novelty may have a role here, the young animal tending to avoid stimuli that are quite distinct from those that it has experienced (Bronson 1968). For example, Coppinger (1970) found that hand-raised blue jays (*Cyanocitta cristata*) avoided novel insects to an extent that depended on their previous experience with similar ones and on the degree of difference between these and that newly presented.

If a food is poisonous but not lethal, avoidance at first sight is not so imperative and other mechanisms may come into play. Garcia *et al.* (1966) showed that rats given food of a particular taste, followed by a drug that made them sick even several hours later, would subsequently avoid that food (see section 6.3.1). This 'Garcia effect' has since been demonstrated in many other species, using a variety of techniques (Gustavson 1977). Although the illness involved can be induced in a number of ways, the feature of the food to which it has proved possible to link this illness has, most often, been taste (Garcia & Hankins 1977). Rats are much less prepared, for example, to learn a connection between illness and the colour of a food or the fact that a buzzer sounds when they are eating it. There may, however, be complications: chicks will learn an aversion to coloured food more readily than to coloured water; on the other hand they will associate water, but not food, with a taste cue (Gillette, Martin & Bellingham 1980). Such mechanisms may lead, in nature, to specific aversions which stop animals eating food that would be harmful to them. A similar system may enable them to distinguish between food and non-food items, although in this case acting through the fact that the former have nutritional consequences whereas the latter do not. When newly hatched, domestic chicks peck indiscriminately at food grains and at sand, but if they ingest the food this leads them to show more pecking at both stimuli about an hour later (Hogan 1973). Only after many hours of experience with both do chicks learn to distinguish between food and sand, so that they peck more at the former. This experience must involve pecking: force-fed chicks do not learn to prefer food (Hogan 1975). The chicks appear to learn that pecking at food alleviates hunger, whereas pecking at sand does not.

Within the spectrum of possible foods, animals frequently prefer some to others and there are several ways in which this may come about. The preferences of some snakes are already present at birth. Burghardt (1967) showed that the newborn young of several species differed in their response, as measured by the number of tongue flicks shown, to extracts of various prey organisms presented on swabs (Fig. 3.6). For example, one species (*Thamnophis radix*) responded especially strongly to extracts of annelid worms but an extract of crickets evoked only the same strength of response as a swab with just water on it. Exactly the opposite was true of *Opheodrys vernalis,* and these differing preferences correspond closely to the natural diets of the two species. Preferences may subsequently change as a result of experience with alternative foods. Burghardt (1971) tested the idea that the initial preferences he had found might stem from the maternal diet; because snakes are viviparous, the mother's food might affect her unborn young. However, the newborn young of the garter snake *Thamnophis sirtalis* showed no difference in their responses to swabs containing worm extract and fish extract according to which of the two diets their mother had been eating. Arnold (1978), who has also carried out extensive experiments on the genetic basis of feeding preferences in garter snakes (see section 1.6.3), has shown that with some prey early experience may have important effects. Young *T. sirtalis* attacked dead fish more readily if they had had a single experience with a live one.

In some other species, the feeding preferences of young animals do seem to depend on the diet of the parent (see Galef 1976). There is, for example, evidence that weanling rats prefer the same diet as that eaten by their mother during lactation and that this preference is passed on by changes in the flavour of the milk (Galef & Sherry 1973; Bronstein *et al.* 1975). More generally, prior

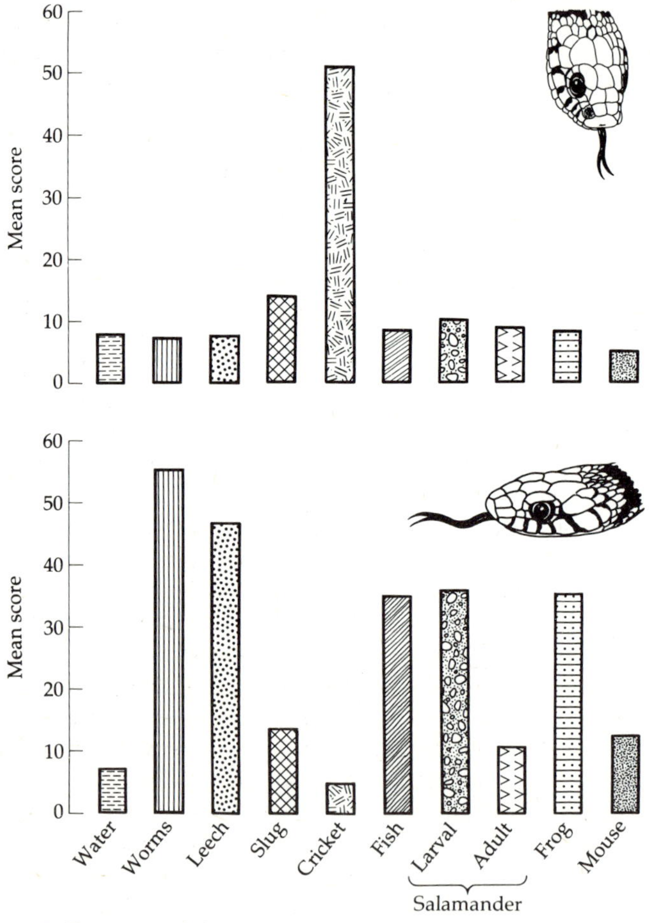

Fig. 3.6. Response of five western smooth green snakes (*Opheodrys vernalis blanchardi*) (above) and of 22 eastern plains garter snakes (*Thamnophis r. radis*) (below) to water extracts from the surfaces of various small animals. None of the snakes had fed previously. The response score is based on the number of tongue flicks and attacks the snakes delivered to each stimulus. (After Burghardt 1967.)

experience of a particular type of food often makes animals more prone to select it when given a choice. In zebra finches (*Taeniopygia guttata*), Rabinowitch (1969) found individuals to prefer a familiar seed type and suggested that this might be partly a preference for foods fed to them as fledglings. As shelling seeds involves skills which differ between seed types, it may also be that experience makes birds more adept at handling the types they have eaten most (Kear 1962). Physiological adaptation leading to more efficient digestion has been suggested by Partridge (1981) as another reason why it may benefit animals to stick to familiar foods.

The most obvious examples where the handling of food involves the development of special skills concern predatory behaviour. In a fascinating, but largely unpublished, study, Norton-Griffiths (1967, 1968, 1969) examined this in the oyster-catcher (*Haematopus ostralegus*). This species is amongst the few wading birds in which the young are fed by the parents, at first near the nest site and later lower on the shore. A probable reason for this is that the species has specialised feeding habits which develop only with experience. Mussels (*Mytilus edulis*) make up a substantial part of the diet and the techniques used by oyster-catchers to penetrate their shells fall into two types, some birds using one, some the other. Hammerers attack closed mussels on dry land, propping them up and then piercing a hole in the shell at its weakest point, on the ventral side. On the other hand, stabbers attack mussels that are underwater with their siphons open, deftly inserting the beak in the right place to cut the adductor muscle that holds the two valves together. By colour-marking birds and cross-fostering, so that chicks were raised by adults using the opposite technique from that of their natural parents, Norton-Griffiths (1968) showed that the young develop the same skill as that of the adults that are rearing them as a result of experience.

In domestic cats (*Felis domesticus*) also, the capture, handling and killing of prey involves complex motor sequences. These actions can be shown by inexperienced animals but experience does have a marked influence upon the behaviour (Caro 1980 a,b). Familiarity with a particular prey species in the first three months of life was followed by increased predation of that prey and improvement in the motor skills employed in dealing with it, but had only limited effects on the level of predation of the cats on other prey types which they had not experienced earlier. The improvement with experience depended to a large degree on whether the mother was present when the prey animals were offered to the kittens. Without their mother present, few of the kittens killed prey; they tended to chase and paw, rather than to bite it as they would if she was there. Thus, the skills used by cats in predatory behaviour are built up by experience during infancy. In this species, as in many other carnivores, mothers often expose their kittens to prey, both dead and alive, and this must have an important influence on the development of their skills. By contrast, Caro (1980b) found no evidence that predatory behaviour was improved by the opportunity to play with inanimate objects during infancy. This could have been simply because of the experimental procedure, the kittens in the group not given such experience inevitably having other objects they could manipulate. Nevertheless, it is an interesting point. The function of object play is, like that of social play (see section 4.3.3), somewhat enigmatic and, because it shares some motor patterns with predation, it has often been thought to provide practice for later killing. Caro's work provides no evidence that this is the case.

As the examples outlined in this section have shown, there is no straightforward generalisation that can be made about how animals come to recognise predators and how they develop both recognition of food and the skills required to handle it. However, we can suggest functional explanations of why natural selection has favoured the different strategies of development. Which of the varied ontogenies that have been described applies to a particular case seems to depend to some extent on the degree to which the aspect of the environment to which the animal must respond is predictable and how imperative it is that the correct response is given on the first occasion that the situation is encountered. The next section describes one of the classic cases of behavioural ontogeny, that of bird song. Despite the fact that every example cited involves learning from other individuals, there are again strong species differences in the details of development which help to illustrate finer points.

3.4 The development of song in birds

Song development in birds is a particularly good topic with which to illustrate the ontogeny of behaviour. It has been extensively studies, particularly since the pioneering work of Thorpe (1958), and has been found to show beautifully how genetic and environmental factors interact during the course of development. It has both sensory and motor components, which are more clearly separable than in many other examples, songs heard being memorised and the bird then matching its motor output to that memory. Because the song is learnt from other individuals, song development also shows the rudiments of culture. Finally, an added bonus in the study of bird song is the existence of the sound spectrograph, an instrument with which patterns of sound may be portrayed visually as plots of frequency against time (sonagrams). This may seem a trivial point, but the detailed study of song necessary if subtle differences are to be revealed would be impossible without the use of sonagrams.

It is probably safe to say that, at least in passerine birds, learning plays a part in the development of all songs (Marler 1981). Simpler sounds, such as the cooing of doves and the crowing of cocks, may develop normally even in birds that have been deafened so that they cannot hear either other individuals or themselves. But in all passerines that have been studied closely the development of song has been abnormal when the bird has been denied the opportunity to hear other individuals singing. Indeed, in some species, simple call notes may also be affected by learning; in the twite (*Acanthis flavirostris*), for example, members of a mated pair have calls which they come to share with each other through learning and which probably help them to maintain contact in a flock (Marler & Mundinger 1975).

The simplest way to ascertain whether or not hearing other birds affects the development of song is to hand-rear young birds in acoustic isolation from the earliest possible age. Ideally, this should be from the egg, but it is often more practicable to start when the chicks are a few days old. The song which results usually has a few rudimentary characteristics typical of the species: its length and frequency range may, for example, be close to those of wild birds. However, the songs of isolates usually differ from normal in many other respects. They may be slower, have fewer elements in them, have an unusual arrangement of elements or be generally more variable. That these defects arise because the young birds have not heard members of their own species singing has been shown by experiments in which tape-recordings have been played to them, either when they start to sing or before that stage, and they have subsequently produced normal song. Indeed this is often a very precise copy of the song presented on the tape (e.g. Slater & Ince 1982; see Fig. 3.7). By contrast, efforts to train young birds in this way with the songs of other species have often been unsuccessful. The white-crowned sparrows (*Zonotrichia leucophrys*) that Marler (1970) trained would learn their own

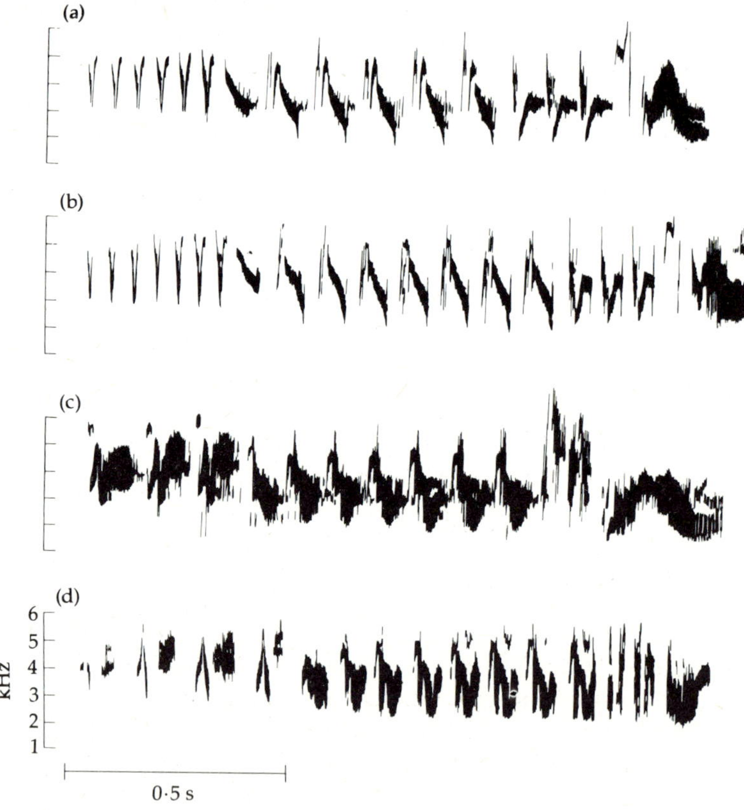

Fig. 3.7. Sonagrams of two songs of chaffinches (*Fringilla coelebs*) played on tape-recordings to a young male of the same species when he was starting to sing himself (tutor songs; a and c), and of the copies of them that he subsequently produced (pupil songs; b and d). Note that copying is generally very accurate but for a few features; for example, in (d) the bird has failed to copy the last part of the terminal flourish of the song in (c). (From Slater & Ince 1982.)

species' song, but not that of a song sparrow (*Melospiza melodia*) or a Harris' sparrow (*Zonotrichia querula*). A remarkable feature here, and also true of some other species, was that the sensitive period for learning, or *memorisation phase,* was over by the hundredth day of life, some time before the start of the *motor phase* when the birds first sing themselves. By detailed monitoring of the sounds they produce, Marler and Peters (1981) have shown that swamp sparrows (*Melospiza georgiana*) can achieve this feat without rehearsing during the intervening months.

Figure 3.8 illustrates some general features of song development following the scheme put forward by Marler (1970, 1976). He suggests that young birds are born with what he refers to as a 'crude template', which gives them a rough idea of what their own species song should sound like. This acts as a filter which is sufficient to prevent them copying any totally inappropriate songs that they hear. As a result of hearing their own species' song

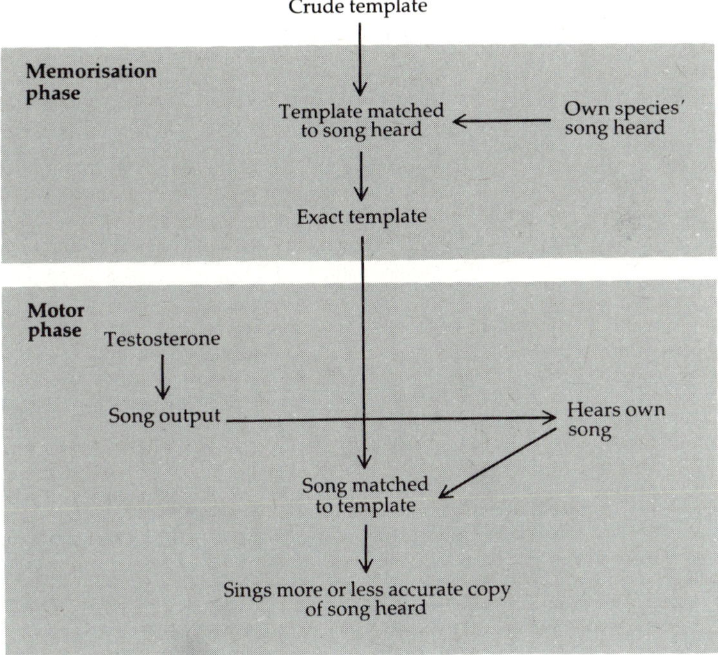

Fig. 3.8. Pattern of song development in young birds able to hear both their own voices and the songs of other birds.

during the memorisation phase, their template becomes sharpened up to become an exact one, a more or less accurate representation of the songs they have heard. This may occur, as in the white-crowned sparrow, before singing starts, or it may be a continuing process so that birds in full song can still modify their output in line with the songs that they hear. Once singing begins, usually in the spring following hatching, when testosterone starts to circulate, the young bird goes through a sequence from subsong, to plastic song, to full song, during which the output becomes progressively louder, less variable and more stereotyped (Marler & Peters 1981). Marler (1976) argues that at this stage the young bird listens to its own output and tries to match it to its template, so that eventually the song that it produces is a near perfect copy.

The effects of isolation and of deafening, illustrated for two North American sparrows in Fig. 3.9, may be put into the same general scheme. Isolation during the memorisation phase (Fig.

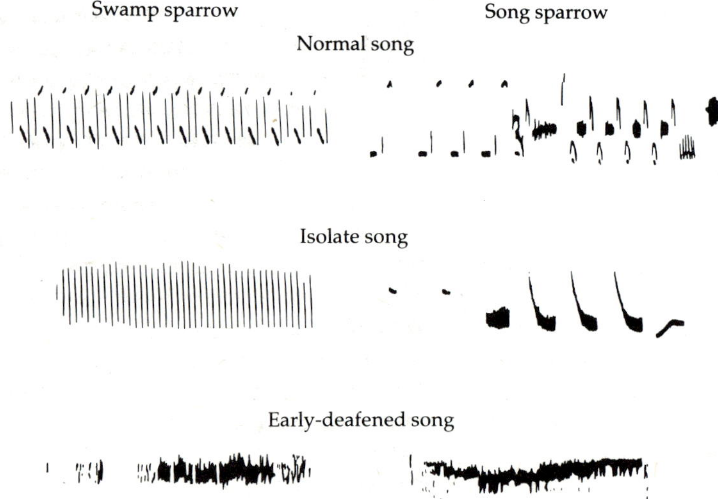

Fig. 3.9. Typical songs of a male swamp sparrow and a song sparrow, shown first in natural form then as developed by males reared in isolation from the song of their species in infancy, and as developed by males deafened in infancy. The isolate songs are much simpler, but still to some degree species-specific, while this is not true of the screech produced by early-deafened birds which have heard neither other individuals nor their own voice. (From Marler 1981.)

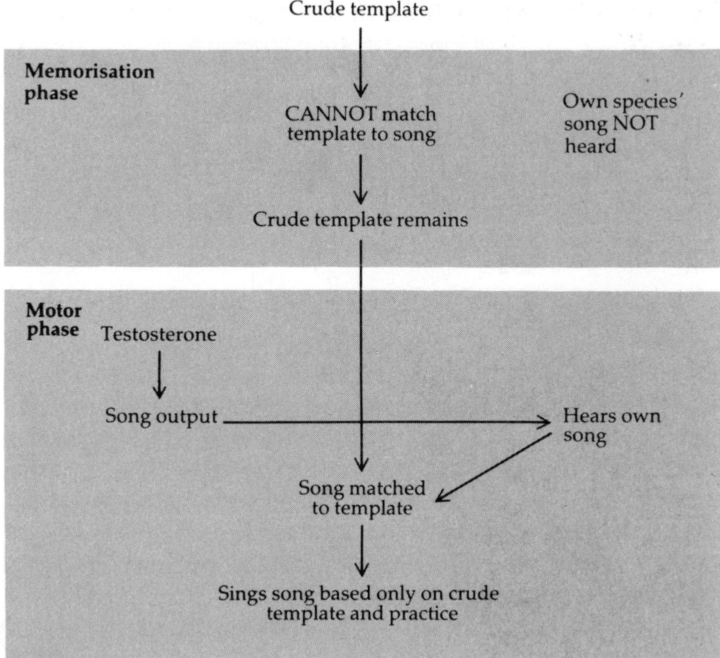

Fig. 3.10. Pattern of song development in young birds isolated as nestlings so that they could hear their own voices but not the songs of other birds.

3.10) renders it impossible for the young bird to make its crude template more exact. All it can do when it starts singing is to match its output to the crude template, and the songs that it produces can only be based on this and on practice. Figure 3.11 illustrates the effects of deafening, here shown as carried out after the memorisation phase is complete. Although the bird has an exact template, it cannot hear its own output, which is therefore inaccessible for comparison with the template. As a result the song is an even more rudimentary effort. Instead, in the extreme case where deafening is carried out very early in life, so that the bird has neither experience of the sounds of others nor even of its own calls, the 'song' may be little more than a screech (Nottebohm 1970).

This general picture proposed by Marler, based largely on his own white-crowned sparrow studies and those of Thorpe (1958) on chaffinches, fits reasonably well for many of the other species that have been studied subsequently. However, some very striking species differences (discussed in greater detail by Slater (in press)) have come to light. These affect particularly the timing of song memorisation, the limitations on what birds are prepared to memorise, and the accuracy with which they learn the sounds that they hear. Each of these will be discussed in turn and reasons suggested why selection may have favoured such species differences.

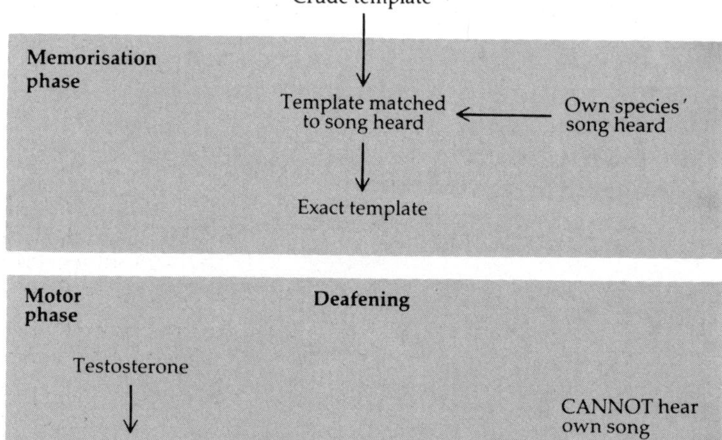

Fig. 3.11. Pattern of song development in young birds deafened after song memorisation so that they could not hear their own voices when they began to sing.

3.4.1 *Timing of memorisation*

The white-crowned sparrow appears to be unusual in having a sensitive period for memorisation which is over well before the bird itself starts to sing. The chaffinch (Slater & Ince 1982) and the marsh wren (*Cistothorus palustris*) (Kroodsma 1978) will also learn very early in life (see Fig. 6.7b); however, they continue to be prepared to memorise songs the following spring, when they start to sing themselves. Interestingly, in the marsh wren the extent to which the bird will learn then depends to some extent on how much it has learned in the previous year (Kroodsma & Pickert 1980). Once a chaffinch is in full song it will not learn any further songs played to it, even if the only song that it has is a rudimentary one typical of an isolate. This suggests that the sensitive period for memorisation ends at a particular time, perhaps when testosterone levels rise, as suggested by the work of Nottebohm (1969), rather than when learning has been completed in any sense. This makes an interesting contrast with imprinting, the classic example of learning involving a sensitive period (see Chapter 4). The evidence suggests that imprinting commences when the animal has reached a particular stage of maturity and terminates when the animal has learnt about familiar objects (Bateson 1979; see Chapter 4). As with much else in development, each stage depends on appropriate predecessors and can go no further than be complete—the upper floor of a two-storey house cannot be built without the lower and the roof has to wait until the upper has been completed. In the case of song learning, however, while a particular stage of maturity may signal the onset of readiness to learn, this readiness also seems to be terminated independently of what has been learnt. Using the same analogy, the roof does not wait for the upper storey beyond a certain stage but instead the house is completed as a bungalow!

In contrast to the species mentioned above, in which some song learning occurs before the young bird sets up its territory, are others in which the primary source of learning is the bird's territorial neighbours. Payne (1981a) has shown this to be the case in indigo buntings (*Passerina cyanea*),where small groups of males on adjacent territories tend to share songs with each other, and captive birds learn songs from individuals with which they can interact socially as young adults. The breeding success of birds which share songs with neighbours is greater than that of birds which do not (Payne, in press), suggesting that the sharing of songs is beneficial, presumably through its role in the social relationships between potential rivals, which may perhaps spend less time fighting when they share songs.

In a final category come species in which learning of songs is not restricted to a clear sensitive period, individuals being able to alter the songs they sing throughout life. Canary song, for example is not fixed in adulthood but the repertoire varies from year to year (Nottebohm & Nottebohm 1978), and Payne (1981b) found that village indigobirds (*Vidua chalybaeta*) changed their songs appropriately when they moved to a new area.

The sensitive period for song memorisation can therefore vary from a short phase early in life, apparently around 40 days long in the white-crowned sparrow, to an unlimited period throughout life. Just to stress the variation between species, it is interesting to note that four species with very different patterns, white-crowned sparrow, chaffinch, indigo bunting and canary, would all at one stage have been classified as belonging to the same family (Fringillidae).

3.4.2 *Limitations on what is learnt*

In the white-crowned sparrow example discussed above, the crude template was seen as a filter excluding the copying of other species, even ones closely related to it. This may not, however, always be true. Even in this species, individuals may learn and pass on to their own offspring the songs of another species with which they are housed (Baptista & Morton 1981). As with the indigo buntings mentioned above, birds may be quite prepared to learn a song from an individual with which they can interact, even if they will not learn it from a tape-recorder. Despite this potential, in the great majority of species, individuals in the wild only sing songs typical of the species to which they belong. Social factors, such as a proneness to learn from particular individuals, for example territorial neighbours or the male that reared them (Immelmann 1969), may to some extent be important in this, but there are certainly also physical limitations on what they will learn, as implied by the 'crude template' idea, which suggests that the limitations are on what the young bird can memorise rather than on what it can produce. It is not easy to get evidence on this, however, for the only way that one can tell whether a bird has memorised a song is when it sings it. But it does seem unlikely that the constraint is a motor one because the syrinxes of different species are very similar, differences between what birds will and will not learn are often very slight, and some species will produce the most amazingly varied songs without having noticeably more complex vocal apparatus than others.

In a study of song development in the swamp sparrow and the song sparrow, two closely related North American species the songs of which are shown in Fig. 3.9, Marler and Peters (1977, 1981) have shown just what it is that constrains young males to learn only the song of their own species. Interestingly, the rules differ between the two. Swamp sparrows will only learn syllables of their own species, but will do so even if these are edited into a sequential pattern typical of song sparrow song. On the other hand, song sparrows are prepared to accept swamp sparrow syllables provided that they are organised in the song sparrow fashion, or song sparrow syllables in the swamp sparrow pattern. They will not, however, learn swamp sparrow song itself. Both species therefore have constraints which are sufficient to stop them learning the song of the other.

In sharp contrast to cases where constraints are strong come some remarkable cases where birds are not at all fussy about what they will learn. In these the filter provided by the crude template must have a very broad mesh indeed and, at least in some cases, there cannot be a requirement for social interactions with the individuals from which learning takes place. Parrots and mynahs in captivity are well-known examples, but perhaps the most striking is a bird whose great breadth of learning is apparent in the field, the marsh warbler (*Acrocephalus palustris*), studied by Dowsett-Lemaire (1979). Adults of this species do not change their songs from one season to the next and the learning appears to occur entirely in the first autumn and winter, when there is no adult marsh warbler song from which to copy. During this period the young birds move south from Europe to East Africa, learning the sounds around them as they go, and when they start to sing the next season each bird mimics on average 76 other species. In all Dowsett-Lemaire has recorded 212 species as being imitated by marsh warblers and it is likely that other sounds in their repertoires are imitations also, but of species she has not herself recorded. Sounds of different species appear in the song in quick succession, jumbled together in an amazing outpouring of variety (see Fig. 3.12). The main constraint on what is produced may well, in this case, simply be a motor one, the bird being unable to reproduce some of the sounds that it hears because they lie outside the frequency-range capabilities of its syrinx.

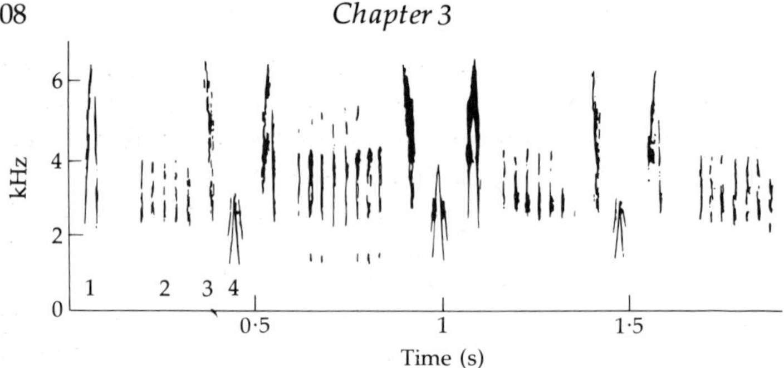

Fig. 3.12. Sonagram of a song of the marsh warbler with complex rhythmic construction alternating four calls of different origin: (1) note of tawny-flanked prinia (*Prinia subflava*), (2) alarm call of Heuglin's robin (*Cossypha heuglini*), (3) call of red bishop (*Euplectes orix*), (4) call of brown-headed tchagra (*Tchagra australis*). (After Dowsett-Lemaire 1979.)

3.4.3 *The accuracy of learning*

Despite the differences in what they will learn, both white-crowned sparrows and marsh warblers learn with great fidelity, so that the sources of sounds they have copied can be determined without difficulty. (Figure 6.7a provides another excellent example of precise copying.) But learning is not always quite so accurate. In the laboratory, birds whose experience is carefully controlled often produce songs which are based upon, but not accurate copies of, those with which they have been trained. This is true, for example, of the song sparrow (Marler 1981) and may explain why there is great variation between the songs of wild males in this species (Harris & Lemon 1972).

The extent to which learning in the wild is inaccurate is not easy to assess because an individual singing an apparently new song may have learnt it accurately from other individuals which the observer has not come across. Nevertheless, inaccurate copying and improvisation certainly do occur during song development in the wild. In wrens (*Troglodytes troglodytes*), for example, Kroodsma (1980) found neighbours to have some sequences of elements in common, but that the overall pattern of song differed between them. Likewise, in chaffinches the different song types found in an area often show similarities not found between songs recorded further apart (Slater & Ince 1979)—a good indication that new song types arise through inaccurate copying, such as has been found in the laboratory (Slater & Ince 1982; see Fig. 3.7). Much of the song learning of chaffinches is extremely accurate, as indicated by the fact that some song types are shared in more or less identical form by many birds in an area, but a population also includes many rarer song types. These must either have been introduced by immigration or created within the population by inaccurate copying. Slater, Ince & Colgan (1980) argued that 15% of the song types in their study area of 42 territories had arisen in these ways, the remaining 85% having been learnt accurately within the population. They suggested that copying errors arising at random were sufficient to account both for the variations they found within the population and for the changes through time in the songs present described by Inch, Slater and Weismann (1980). As inaccurate copying is rather rare it seems unlikely that selection favours individuals showing it; it probably arises simply because some birds have fewer learning opportunities than others.

3.4.4 *Why does song development vary so much?*

From this brief survey of bird song development we can get an idea of the factors that are important if it is to take place normally. As with many other developmental processes, there tends to be a sensitive period during which young birds are much more prone to memorise sounds that they hear than at other times. Within that sensitive period the sounds that are learnt tend to lie in a limited range of those that are heard; there are undoubtedly species-specific constraints and, in some species, social relationships may make it more likely that one bird will learn from another particular individual. Other factors important in shaping the finished product are the extent to which young birds improvise rather than copying accurately, the extent to which adults sing during the sensitive period and so provide opportunities for copying, and whether or not song learning and song production are affected by testosterone. Clearly, in each species, there are many different factors involved.

The sorts of differences in development between species described above must have arisen for good functional reasons; they are too great to be thought of as different means to the same end. All that can be done at this stage is to provide hypotheses as to why natural selection may have led to such differences. Similar suggestions have been made independently by Catchpole (1982) and by Slater (in press) and seem to provide a reasonably plausible picture.

Of the various functions suggested for song two seem predominant: that it acts as a mate attractant and that it acts as a repellent to other males. The hypothesis suggests that one of these is more important in some species and the other in other species, and that this leads to the song and its development being adapted to one role rather than to the other. Songs which are primarily mate-attractant are likely to have been influenced by sexual selection, leading to the very large number of different phrases that some birds use. The stress here would be on enhancing attractiveness by variation, which could be achieved either by improvisation or by a wide variety of sounds being copied accurately. There would be no strong reason for the variety of sounds to be copied to be limited, or for the copying to be restricted to a short sensitive period. On the other hand, where song acts mainly as a signal between males, birds on neighbouring territories often have just one or a few different phrases which are shared between them and if one sings a particular phrase, others will tend to counter-sing with it. For this sort of system to operate, very accurate learning is required, the young bird must learn either from particular individuals or the exact sounds to be matched, and the learning process must be completed early enough for the bird to use the songs as he sets up his territory. According to this hypothesis there are two different reasons why learning is involved in the song development of all passerine birds. Where mate attraction is an especially important function, learning may have evolved as a means to achieve variety; where song is more important as a signal between neighbours, learning may be the only means whereby young males, which cannot predict where they will find a territory, can match their neighbours accurately wherever they settle.

These ideas are obviously speculative and not very easy to test. But they do illustrate the point that what might seem to be minor differences between species in some aspect of their way of life can have ramifying consequences for many different aspects of behaviour and its development.

3.5 Conclusion

In this chapter various aspects of behavioural development have been illustrated by taking a few well-researched examples which show how both perceptual and motor abilities arise during the lifetime of the individual. Here we will summarise the variety of ontogenies and consider some of their implications.

At one end of the spectrum are behaviour patterns which appear more or less fully formed the first time that the young animal finds itself in the appropriate situation. Pied flycatchers mob shrikes without ever having seen one; in contrast to the young of ground-nesting species, young of the wood duck (*Aix sponsa*) show no fear of the steep side of a visual cliff. It is important to realise that such capabilities do not imply that the behaviour does not develop; it will not appear before a particular age, so that some degree of maturation is certainly required, but many other aspects of the interaction between the animal and its environment, even including learning in earlier situations, may be involved. In several other examples, the behaviour appears when first required but subsequent experience also has a marked effect upon it. Goslings will show alarm in response to a hawk when they first see one, and it seems that a stimulus of this form is especially effective. However, they also respond to many other stimuli and only slowly do they come to ignore those that they see often and which are harmless. Being carnivores, hawks are rare, and it may simply be that they are not seen often enough for habituation to set in. The pecking of gull chicks seems to follow a similar pattern. A rather vague predisposition channels their interest towards the parent's beak and, as a result of experience, perhaps in this case supplemented by reinforcement, leads them to a very precise knowledge of what should be pecked at.

The beak of the adult laughing gull is a highly predictable stimulus and chicks must peck at it to get food. Their responsiveness when they hatch is perfectly adequate to ensure this. But what of later in life? Some species do concentrate on a few food items but, if the food supply is unpredictable, the capacity to change diet in lean times is essential. Herring gulls, for example, may feed on crabs and mussels along the shore, on earthworms in fields, or by scavenging on rubbish dumps (McFarland 1977). How can they know what to eat? They may have some preferences without the need for experience, like garter snakes, and their parents may introduce them to particular food types, like oystercatchers. But, with such a catholic diet, learning from individual experience probably has an important part to play. Like rats, they may eat a little of something new and later avoid it if it makes them sick or, like chicks, they may continue to eat foods that alleviate hunger and have good nutritional consequences.

Bird song is another topic in which we have argued that learning plays a role because the environment is unpredictable. If song varies and the bird must match its neighbour, it could not do it without copying. In this case there is the added complication, and interest, that copying is from other individuals. Song therefore changes from place to place and from time to time, just like human languages, and is an example of culture—the passing of information from one individual to another through copying. We have considered several other examples of this, blackbirds learning what to mob and oystercatchers learning how to eat mussels being two. There have also been some fascinating cases documented in wild animals where one individual has developed a skill which has then spread through a population. That curse of those who like cream on their porridge, the milk bottle top opening of tits, spread rapidly through Britain between the wars by cultural transmission (see section 2.3). The washing of sweet potatoes before eating them, and various other skills, first developed by a particularly inventive Japanese monkey (*Macaca fuscata*) called Imo, spread by imitation to nearly all the animals in her troop (Miyadi 1964). These are, however, rather isolated examples. It is in humans that the cultural transmission of information has really flowered; once Einstein had described the laws of relativity there was no need for anyone else to discover them anew. Learning from other individuals has tremendous potential as a means of adapting animals to a rapidly changing world or one that varies greatly from place to place, but these features are more true for humans than they are for other species. Humans also make use of tuition, the production of a behaviour specifically as an example to another individual, as well as imitation. This is not easy to demonstrate in other animals. How could one show that singing of a bird late in the season was specifically so that his offspring could learn the song, rather than for any other reason? Nevertheless, there are certainly a few examples. As often happens, the best must be the honeybee (*Apis mellifera*), where the waggle dance of returning foragers transmits to other animals the distance and direction of a food source (von Frisch 1974). Not only does this seem a reasonable example of tuition, but it is carried out in a symbolic language too!

3.6 Selected reading

Bateson's articles (1976a, b) make a number of important general points about the subtleties of behavioural development, and the study by Hailman (1969) provides a most convincing and rewarding case history. The book edited by Pick and Walk (1978) has useful contributions on visual system development. Galef (1976) reviews cultural transmission, including that of food preferences and predator avoidance. A more general and highly readable review of cultural transmission is the book by Bonner (1981). The most recent reviews of the bird song learning literature are a brief paper by Marler (1981) and a more extensive account by Slater (in press).

CHAPTER 4 THE DEVELOPMENT OF SOCIAL RELATIONSHIPS

NEIL CHALMERS

4.1 The nature of social relationships

When animals react to each other as individuals, rather than simply as members of the same species, and when their behaviour towards each other consists of more than a series of independent, isolated interactions, they can be said to have a social relationship. Social relationships are rarely static over time, and this chapter describes some of the developmental changes that occur in them, some of the consequences of these developmental changes, their possible adaptive significance and some of the causal mechanisms that bring them about.

Although the term 'social relationship' is common in everyday speech, it has acquired a more specialised and clearly defined meaning in the behavioural sciences. Hinde (1979), in particular, has identified its definitive features. First, a relationship can only be said to exist between two individuals if those individuals continue to interact with one another over an extended period of time. A mammalian mother and her infant, and a mated pair of monogamous birds such as swans, both provide examples. Secondly, the individual identity of the animals involved in the relationship is important. Two worker honeybees from the same hive, or two fish in the same school, would not usually be thought to have a social relationship with one another. They may interact with one another intermittently over an extended period of time but purely as members of the same hive or school, rather than as specific individuals. The nature of their interactions may be no different from those in which they take part with any other member of the hive or school.

Although the interactions between two individuals may be intermittent, the relationship that exists between them is nonetheless continuous. If a female herring gull *(Larus argentatus)* leaves her nest and mate to collect food, her relationship with her mate survives her temporary absence. A relationship therefore transcends the moment-to-moment interactions and behavioural adjustments of which it is composed. The collective properties of these interactions and behavioural adjustments constitute the relationship. In Hinde's terminology, the study of a relationship involves an analysis of the patterning in time of the constituent interactions together with an analysis of the content and the quality of those interactions.

The vocabulary that is needed to describe the various attributes of a relationship is different from that required to describe the properties of its component interactions, and the kinds of analysis that are required to understand the developmental changes in a relationship are frequently different from those required to understand the development of particular kinds of interactions.

4.2 Changes in relationships during development

Although most social relationships change over time, the changes that occur in the relationships between a young animal and those that care for it as it grows older are particularly striking. Baby baboons (*Papio anubis*), for example, spend much of their first week or two of life clinging to their mother (Altmann 1980; Chalmers 1980b), but as they grow older, they spend increasing amounts of time away from her. In the Barbary or ring dove (*Streptopelia risoria*), parents regurgitate food for their young during the first 15 days after hatching, but after that refuse to do so when their young beg for it. The young are then forced to peck at grain and other objects, and so come to feed independently.

Developmental changes in relationships such as these prompt three questions.

(i) What conditions are necessary if the young animal is to form a relationship with its mother in the first place?

(ii) What part is played respectively by the mother and by the young animal in bringing about the changes in their relationship with age?

(iii) Does the nature of the relationship between an infant and its mother have long-lasting consequences for its development?

Each of these questions is framed here in terms of the relationship of a young animal with its mother, but similar questions can be asked about an infant's relationship with other animals in its social group such as its father, its brothers and its sisters. The remainder of this section is devoted to the above three questions concerning mother–infant relationships. Section 4.3 discusses the development of the relationships of infants with their fathers and age mates.

4.2.1 *Initial formation of a relationship*

Many animals when born or hatched are virtually helpless. Mouse and rat pups are naked, blind and immobile, as are many newly hatched birds. Baby monkeys and apes are able to cling tightly to their caregivers and can suck milk from their mother's nipple, but otherwise can do very little for themselves. Species with young born in this condition are said to be *altricial*. In altricial species the mother must play an active part in establishing a relationship with her infants. If she is to do this, it is important that the young present the correct stimuli. This is shown particularly clearly in an experiment carried out by Noirot (1964, 1965) on laboratory mice. Adult female mice lick 1-day-old pups and carry them back to the nest if the experimenter places the pups some distance away. The females are much less likely to respond, however, if the pups are dead. Noirot's experiments suggest that one of the most important stimuli that elicits maternal behaviour is the ultrasonic vocalisation which the live pup emits. Adult females can be made more likely to behave maternally to dead pups by previous exposure to living pups that they could hear (and smell).

The appearance and behaviour of human babies also affect their attractiveness to adults. In an experiment carried out by Sternglanz *et al.* (1977), adults of both sexes were presented with drawings of babies' faces that varied systematically in such features as the width and height of the eyes, the distance of the eyes from the top of the head and the size of the pupils. The babies that proved the most attractive were those with large eyes and a large forehead (Fig. 4.1). Smiling, sucking the mother's nipple, and wobbly, uncoordinated movements also make the baby attractive to the mother.

A mother's readiness to behave maternally towards newborn animals is not only influenced by stimuli emanating from the

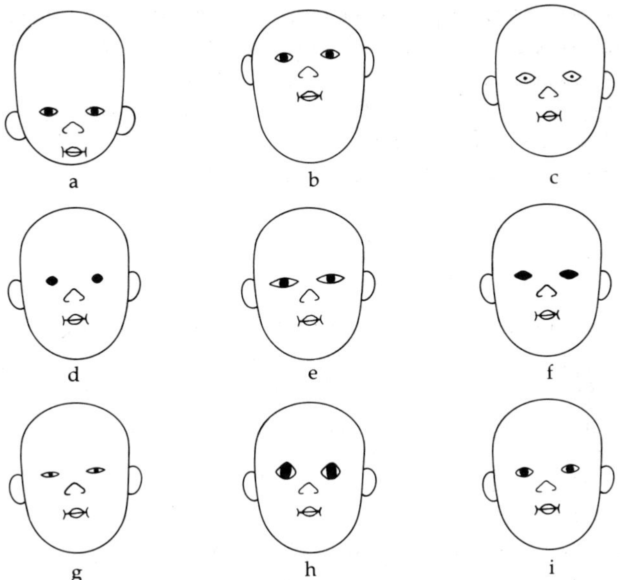

Fig. 4.1. Drawings of babies' faces that varied systematically in several parameters, and were presented to adult observers. A combination of high forehead, large eyes and large irises (h) proved to be most attractive. (From Sternglanz, Gray & Murakami 1977.)

young, but also by her own reproductive condition and her previous experience of raising infants. For example, a virgin rat shows little maternal behaviour if she is confronted with newborn pups, but a rat that has just given birth immediately starts to lick and nurse them (Rosenblatt & Lehrman 1963). In sheep also, a ewe's behaviour towards her firstborn lamb differs from her behaviour towards subsequent lambs (Poindron & Le Neindre 1980). Mothers of firstborn lambs frequently fail to stand up when the lamb tries to reach the udder, they delay licking their lambs for longer after birth than do experienced mothers, and they are more likely to butt their lambs aggressively.

The behaviour that mammalian and avian mothers show towards their altricial young during the early stages of their relationship includes feeding, cleaning, thermoregulation, either by cuddling or shading the young, and protection against predators and other sources of danger. In addition, particularly in the great apes and human beings, the mother engages in long and intense interactions with her baby which serve no immediate physical need and yet seem to form a vital part of the developing relationship. Chimpanzee (*Pan troglodytes*) mothers, for example, spend long periods of time fingering and gazing at their infants and dangling them in the air (van Lawick-Goodall 1968). Human mothers talk to their babies in a characteristically slow, exaggerated and repetitive manner and, while she does so, mother and baby may gaze at each other for unbroken periods of many seconds (Snow 1972). Mothers report that episodes like this make them feel bound much more closely to their babies.

Although altricial young cannot play a very active part immediately after birth in the mother–infant relationship, and although their behaviour initially consists of little more than automatic responses to certain situations, such as protesting when hungry or cold, this state of affairs does not last for long. The young animal, and in particular the young human baby, rapidly learns the consequences of its own and its mother's behaviour, and starts to interact actively with its mother. Indeed, within a few hours of birth the human baby moves its arms and fingers in synchrony with its mother's utterances, and within two months it has developed a unique style of 'conversation' with her (Bower 1979; Trevarthen 1975).

Similarly, baby baboons in their first weeks of life emit screams and squeaks indiscriminately when frightened by either animate or inanimate objects, and these sounds bring the mother rapidly to the baby's side if she was previously absent. After a few weeks, however, the infant becomes more selective about the situations in which it screams and squeals, and comes only to emit these noises, and so elicit the mother's aid, when interacting with individuals in the group (Chalmers 1980a).

In altricial species, therefore, the mother–infant relationship starts out with the mother adopting an active, controlling role in the relationship and the infant responding to the mother with a small set of automatic, reflex-like responses. The nature of the relationship rapidly changes, however, as the infant learns to adjust its own behaviour to its mother's and learns some of the social consequences of its own actions.

Some birds and mammals are born in a much more advanced state of development than the young of altricial species. From the outset their sense organs are well developed, and their body surface is insulated by down or hair. They are also able to stand and move freely within the first hours of life. Such species are said to be *precocial*. The mothers of precocial young may themselves be highly mobile after giving birth. Wildebeest mothers, for example, give birth in northern Tanzania during a long migration (Estes & Estes 1979). This means that, unlike altricial species, young precocial animals must actively follow and stay close to their mothers if they are to survive. That is, both the mother and her offspring must play an active role in establishing their relationship.

Precocial young normally become attached to the mother and follow her within the first day or two of life. They subsequently learn not to follow, and indeed to avoid, objects and individuals that do not resemble the mother. This phenomenon is called imprinting or, more precisely, *filial imprinting*, to distinguish it from sexual imprinting, which is discussed in more detail below. Laboratory experiments on young precocial birds such as ducklings, goslings and chicks have shown that their responsiveness to objects changes markedly with age. There is a short period, often termed the sensitive period, early in such animals' lives, when they will follow a wide variety of objects to which they are exposed (for reviews, see Bateson 1966; Sluckin 1972). Before the sensitive period they will not follow anything, and after it they will only follow an object with which they have become familiar during the sensitive period. The range of stimuli that a chick, duckling or gosling will follow during the sensitive period is very wide and includes balloons, buckets and human beings. However, some objects are more effective than others. Particularly favoured are objects which are conspicuous, such as flashing lights and boldly patterned boxes, but which do not move too violently. Auditory as well as visual stimuli can play an important part in imprinting. This has been demonstrated especially clearly in ducks. Ducklings of many species will, for example, approach rhythmic sounds, and numerous investigators have commented on the powerful, attractive effect of mother ducks' calls on their ducklings (Ramsay 1951; Gottlieb 1974).

Imprinting is a very striking phenomenon and has attracted a great deal of attention, particularly as it raises a number of extremely interesting questions about the development of social behaviour. First, there are questions about the mechanisms that control the young animal's responsiveness to objects and animals that it meets. What determines the onset and termination of the sensitive period? What is the nature of the learning process involved in imprinting? Is imprinting unique, or is it a particularly clear instance of the widespread phenomenon in behavioural development that animals are especially sensitive to their environment at certain periods? Secondly, there are questions about the adaptive significance of imprinting. What advantage, if any, is there for a young animal to learn its mother's features over a short period early in life? Why does the young animal's selective following of its mother depend upon learning for its development? A detailed discussion of these questions lies outside the scope of this chapter, but several of them are dealt with briefly below (for a fuller discussion see Bateson 1966, 1979; Immelmann 1972, 1975).

Various experimental procedures have been used to investigate imprinting, almost all of them using young precocial birds. In most experiments the young bird is placed in an arena or some other enclosure, and is presented with one or more objects (see Fig. 4.2). In one experiment, carried out by Hess (1959), chicks of various ages were put singly into a circular arena with an object that was totally novel to them. The object was moved mechanically in a circle around the arena and the extent to which each chick followed it was measured. Hess found that chicks would only follow novel objects from about 5 to 24 hours after hatching. In another experimental arrangement the young birds are presented simultaneously with two objects, and their preference for one or the other can be examined. In some experiments both objects are entirely novel to the birds, but in others the birds have previously been exposed to one of them, usually for a short period such as half an hour, and the effect of this previous exposure on the birds' preferences can thereby be investigated.

It is now clear that, although there is a definite period when young precocial birds will approach and follow novel objects, the duration of this period is quite flexible. Its beginning seems to depend upon several factors. In the first few hours after birth, chicks are not very mobile and so the first appearance of the young bird's following response must necessarily depend upon the development of its mobility. Also, if a young bird is to respond to an object it must obviously be able to detect it, and hence the onset of the sensitive period must depend upon the visual, auditory and possibly other sensory systems being developed to a sufficiently advanced stage; the bird must have reached a certain age from the point of conception. The beginning of the sensitive period also depends on the number of hours that have elapsed since hatching. The end of the sensitive period is influenced strongly by the young bird's experience during this time, and a key factor here is familiarity. As a bird becomes familiar with one object, so this promotes in it a tendency to avoid objects with which it is unfamiliar. However, a bird that is forced to live after the end of the sensitive period with an object that it initially avoids may eventually come to accept it, and even to prefer it to a previously favoured object (see section 2.8 for a fuller discussion of factors affecting the timing of the sensitive period for filial imprinting).

The adaptive significance of filial imprinting would at first sight seem to be clear. A young bird that learns to follow its mother would seem to be much more likely to survive than one that did not. But this does not explain why young birds should have to

Fig. 4.2. A typical test arena used in an imprinting experiment together with stuffed adult female pintail, mallard and redhead ducks used as stimulus objects. In the arena a duckling is shown with a stuffed mallard during a training session. When subsequently testing the duckling's preference between the familiar and an unfamiliar female, the unfamiliar female is suspended from the opposite end of the T-bar. (From Johnston & Gottlieb 1981.)

learn the characteristics of their mother. Why has natural selection not acted to produce young birds that innately follow objects that have visual and auditory features unique to adult females of their species? The answer may be, as Bateson (1979) has suggested, that during imprinting a duckling or chick is becoming selectively responsive not to the generalised appearance of adult females of its species, but to the individual characteristics of its own mother. This could be important because adult female birds of many species will ignore or even attack young birds that are not from their own brood. It is hard to see how any mechanism except one that involved learning could achieve this selective responsiveness towards an individual.

Although this explanation is plausible, it is important to discover whether young precocial birds do indeed respond selectively to the individual characteristics of one specific female, the mother, or whether they will follow any adult female from the same species. An experiment by Johnston and Gottlieb (1981) suggests that mallard ducklings are not very good at distinguishing adult female ducks that they had earlier encountered during the sensitive period from unfamiliar female ducks of closely related species. In their experiment, Johnston and Gottlieb exposed 24-hour-old mallard ducklings for 20 minutes to a stuffed female of one of three sympatric duck species: the mallard itself (*Anas platyrhynchos*), the redhead (*Aythya americana*) and the pintail (*Anas acuta*) (Fig. 4.2). Later, the ducklings were exposed to a pair of these stuffed females, one of which was the female it had already encountered for the 20-minute period and the other of which was completely unfamiliar. In all except one instance, the ducklings showed no preference for the familiar over the unfamiliar stuffed female. The appearance of females of different species is likely to be more diverse than that of females from the same species, and this experiment, taken at face value, suggests that imprinting may not result in young mallards showing preference for one individual female of their own species over another. However, experiments that fail to reveal predicted differences can be difficult to interpret, and one cannot easily dismiss the alternative explanation that young mallards can indeed discriminate between alternative adult females but that Johnston and Gottlieb's experiment failed to reveal this ability. Scott and Bateson (in preparation) have shown that young Japanese quail

(*Coturnix coturnix*) can discriminate between individual adult females in the laboratory, and it is probable that they can also do so in the wild.

Despite the large number of laboratory experiments on imprinting, there has been comparatively little investigation of the formation of mother–infant attachments in wild precocial birds. Without such studies, of course, adaptive explanations of imprinting are difficult to assess. Field observations suggest, however, that it is not simply the young birds that are responsible for maintaining contact with their mothers, and that filial imprinting is not the only mechanism by which early attachments are formed. Indeed, the mother can play a large part in maintaining contact with her brood. Mother ducks of many species may often become separated from their broods for short periods of time, and it is often the mother rather than the young which re-establishes contact (Stewart 1958, 1974; Beard 1964). If the adaptive significance of imprinting is to be understood more fully, therefore, it will be important to discover the relative contribution that it makes towards the initial attachment between the young animal and its mother under natural conditions, to investigate further the extent to which imprinting involves individual recognition, and to determine the extent to which young precocial birds direct their following responses to fellow brood members, thereby producing group cohesion.

This discussion of filial imprinting has concentrated so far almost entirely on birds. Nonetheless, precocial mammals such as goats, sheep and various antelopes also show a strong tendency to follow their mothers shortly after birth and appear to restrict as a result the range of animals to which they will later respond. Mammals appear to differ from birds, however, in certain aspects of the way in which attachments between mothers and infants are formed. For example, in mammals with precocial young, the mother tends to develop an attachment to her young more rapidly than they develop an attachment to her. Moreover, the development of the mother's attachment depends to a large extent on smell. This has been shown, for example, in sheep (Poindron & Le Neindre 1980). Ewes lick their newborn lambs, and stay close to them during the first hours after birth. Normally a ewe will only suckle her own lamb and will reject any other lambs that try to suck from her. However, ewes can be made anosmic (artificially made unable to smell), and if this is done, the ewe will allow both her own and strange lambs to suck from her. Thus, the mother must be able to perceive the lamb's smell if she is to be able to discriminate between it and others. Poindron and Le Neindre produce evidence suggesting that not only does the lamb's smell promote the mother's special attachment to her own infant, but that it also helps to maintain the mother's overall maternal responsiveness. Ewes that cannot smell their lambs tend not only to be undiscriminating in their maternal behaviour, but also less responsive towards lambs generally.

The fact that filial imprinting is found in both mammals and birds makes one ask whether it is fundamentally the same process in the two groups. This question can be put in a broader context because it is quite clear from an increasing number of studies that sensitive periods are found at various stages in the behavioural development of a wide range of species. Dogs (*Canis familiaris*) and rhesus monkeys (*Macaca mulatta*), for example, both have periods in their development when they are exceptionally vulnerable to social isolation. One can therefore ask whether sensitive periods of the kind that are observed in filial imprinting, as well as in other aspects of behavioural development, depend upon common underlying processes. It is premature to offer anything other than 'probably not' as the answer to this question. That an animal varies in its responsiveness to its environment at different stages of its development is scarcely surprising, and there seems to be no compelling reason why one should postulate a common underlying process. The sensitive period observed in young precocial birds in studies of filial imprinting is affected by a multitude of factors both internal and external to the young animal. It seems unlikely that many of these factors will be the same as those that affect the sensitive periods that other animals go through at other stages in their lives with respect to other aspects of behavioural development.

4.2.2 *Responsibility for changes in the mother–infant relationship with age*

At first sight it might seem that the infant's increasing independence from its mother with age stems mostly, if not entirely, from changes in its own behaviour. After all, the infant daily grows bigger, stronger, better coordinated and more mobile, and these changes would seem automatically to promote its independence. There is now ample evidence, however, that females of many species play an important part in advancing the independence of their offspring.

An important study demonstrating the mother's role in encouraging the independence of her young one was carried out by Hinde and his colleagues on captive rhesus monkeys (Hinde 1974). The animals were kept in small social groups consisting of a single adult male, a few adult females and their young. Various aspects of both mother and infant behaviour in relation to each other were measured at regular intervals (Fig. 4.3). Hinde found that the proportion of time that infants spent both off their mothers and at a distance from them out of their immediate reach, increased with age. Several lines of evidence pointed to the mother playing an important part in these changes. First, the mother rejected an increasing proportion of her infant's attempts to climb on to her. Secondly, episodes in which the mother and her infant were in physical contact with one another were initiated by the infant on a larger percentage of occasions with increasing age, and were terminated by the infant on a smaller percentage of occasions. A single index can be obtained from these two percentages by subtracting the latter from the former. Positive values of this index indicate that the percentage of contacts initiated by the infant exceeds the percentage of contacts terminated by the infant; negative values of the index indicate the reverse. A change from negative to positive values with increasing age (see Fig. 4.3) thus indicates that the infant plays an increasing role, and the mother a diminishing one, in maintaining physical contact between the two.

The difference between being in contact with the mother and out of contact with her is undoubtedly of great significance to the infant, but so also is the difference between being close to the mother and at a sufficient distance from her to be out of her immediate reach. If an infant is more than about 60 cm from its mother, the mother has to get up and move to it if she is to take the initiative in helping it. An index similar to that described above can therefore be constructed to analyse the part that the mother and infant, respectively, play in regulating the distance between them, and this provides the third piece of evidence of a change in maternal behaviour. Figure 4.3 shows the percentage of approaches (from

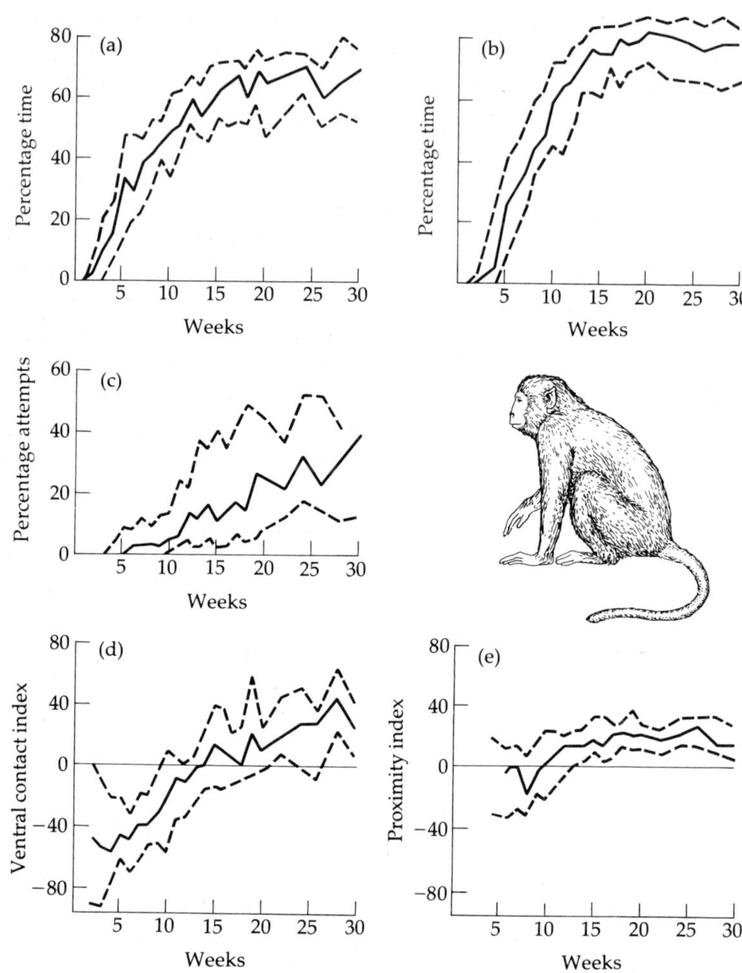

Fig. 4.3. Changes in behaviour between mother and infant rhesus monkeys (*Macaca mulatta*) with age. (a) Percentage of time that the infant spends off its mother. (b) Percentage of time that the infant spends more than 60 cm from its mother. (c) Percentage of attempts by infant to cling ventrally to its mother rejected by the mother. (d) The relative contributions of the mother and infant in maintaining ventral contact. The higher the index, the greater is the part played by the infant in maintaining contact. (e) The relative contributions of the mother and infant is maintaining proximity. The higher the index, the greater is the part played by the infant in maintaining proximity. The unbroken lines represent medians, the dashed lines the interquartile range. (From Hinde 1974.)

further than 60 cm to nearer than 60 cm) initiated by the infant minus the percentage of separations initiated by the infant. Again, the mother plays a decreasing role and the infant an increasing role in maintaining proximity between them.

Studies of wild primates by other workers lead to similar conclusions. Within as little as a day or two of giving birth, olive baboon mothers may place their infants on the ground and walk away from them, forcing the infants to move for themselves (Ransom & Rowell 1973). Altmann (1980), working on yellow baboons (*Papio cynocephalus*) in the Amboseli National Park, Kenya, showed that in the first two to three months of the infant's life the mother decreased the distance between herself and her infant more than she increased it, whereas after three months the reverse was the case. Moreover, the mother's demeanour changed as her infant grew older. In the early weeks of its life, she would constantly watch her infant while it moved around her, and would immediately pick it up if it gave a distress cry. By contrast, mothers of older infants paid little overt attention to their offspring, would move off without them, and would ignore any distress cries that such a departure might provoke.

It is not only in primates that the mother plays an important part in promoting the infant's independence. In ungulates, rodents and carnivores, the mother can frequently take part in driving the young away, especially after she has given birth to her next offspring. In many species, there can be a considerable conflict between the mother and her offspring at the time of weaning, the mothers promoting the independence of their infants more rapidly than they are seeking it. The widespread occurrence of weaning conflicts raises some extremely interesting theoretical questions that were first explored in detail by Trivers (1974). One might have expected that natural selection would have led to a harmonious situation in which at each age the mother promoted the infant's independence to the same degree that the infant sought it. Trivers was able to show that natural selection would not be expected to lead to this state of affairs, and that there should be a time during development at which the infant seeks more care than the mother is prepared to give. His reasoning was as follows. A mother both derives benefits and incurs costs by caring for an infant. She increases its chance of surviving and growing to reproductive maturity, and so she benefits by the increased number of fertile progeny she is likely to leave. Raising infants is physically demanding and very time-consuming, however, so by committing resources to rearing a particular infant or family, the mother is reducing her chance of successfully rearing infants in the future. When an infant is born, the benefit to the mother of looking after it will exceed the cost to her. As it grows bigger, however, the cost will increase, especially because the baby will demand more food. Moreover, the cost of *not* looking after the baby will decrease because the infant will become more self-sufficient. There comes a time, therefore, when it is to the mother's advantage to stop caring for her infant and to start a new family.

The situation is different from the infant's point of view. The cost to it of receiving the mother's care is different from and less than the cost to the mother. By caring for her current infant, the mother reduces the chance of her successfully rearing future offspring; by demanding care from its mother, an infant reduces the chance of its acquiring future brothers and sisters. To the mother, both her current and her future offspring are equally valuable in that they are equally likely to be carrying copies of her own genes. To the infant, however, its own survival is more important than the survival of its brothers and sisters. Genetically speaking, it is completely related to itself, but only a fraction of its genes are identical copies of those found in its siblings by virtue of their having the same mother. Suppose that, at a certain time after giving birth, the cost to the mother of caring for the infant outweighs the benefit that she receives, and that she stops caring for her infant at that time. The cost to the infant of continuing to receive care will *not* be greater than the benefit it receives at that time and so it will continue to seek care, and there will therefore be conflict between the mother and her infant. This will only cease when the infant reaches an age at which the cost of receiving care exceeds the benefits. At this point the infant will stop seeking care from its mother.

It is interesting to see what happens if some of the assumptions in Trivers' explanation are changed. For example, there may be a time in a mother's life when caring for one infant actually increases the chance of her future offspring surviving. As mentioned earlier in the chapter, inexperienced mothers may provide poorer infant care than experienced ones. In such instances one might expect the cost to the mother of raising her firstborn to be relatively low, and hence, other things being equal, less mother–infant conflict might be expected between firstborn infants and their mothers than between experienced mothers and their offspring. Equally, there may be circumstances where not all of a mother's future offspring are of equal value to her. If the chance of the mother surviving beyond the weaning of her current infant diminishes markedly at a particular age, then her current infant would become more valuable to her relative to future ones, and mother–infant conflict would be expected to diminish. Trivers' hypothesis is therefore important in that, if correct, it provides a general explanation for a widespread phenomenon in social development, and in that it generates several interesting and testable predictions about social development.

4.2.3 *Long-term consequences of early relationships*

Sexual imprinting

There is now substantial evidence that the way in which an infant develops, and indeed its very survival, may depend upon its early social experience, and in particular upon the nature of its relationship with its mother. One of the most striking examples of the long-term effects of an infant's early social life is found in the phenomenon of sexual imprinting. This leads a mature animal to direct its sexual behaviour preferentially towards individuals similar to those which it encountered when it was young. Sexual imprinting is most clearly revealed in cross-fostering experiments. In these, the experimental animal, usually a bird, is reared by parents belonging to a different species. After the young bird has reached sexual maturity it is closeted with two birds of the opposite sex, one belonging to its own species and the other to the species which reared it. The bird is said to be sexually imprinted if it directs its sexual behaviour towards the partner belonging to the foster species rather than to the conspecific bird. Sexual imprinting has been studied particularly thoroughly in estrildine finches. Two species from this group are the zebra finch (*Taeniopygia guttata*) and the Bengalese finch (*Lonchura striata*). If a male zebra finch is raised by Bengalese finches, and at maturity is placed with a female zebra finch and a female Bengalese finch, he will direct his sexual behaviour almost exclusively towards the Bengalese female, even though the zebra finch female herself directs sexual behaviour towards him, whereas the Bengalese female does not (Immelmann 1972).

Sexual imprinting occurs in numerous bird species, ranging from herons to parrots, and may also occur in some mammals, although the evidence that mammals direct specifically sexual behaviour at the type of individual with which they were reared is not very clear (Immelmann 1972). As with filial imprinting, sexual imprinting raises numerous extremely interesting questions. It is particularly instructive to compare the two phenomena because there appear to be certain striking differences between them, as well as some similarities. Thus, in both there is a sensitive period during which the animal is predisposed to form an attachment to another individual. However, the sensitive period for sexual imprinting typically occurs later in development than that for filial imprinting. Whereas filial imprinting usually occurs within the first days of life, the sensitive period for sexual imprinting, in ducks and geese for example, does not begin until some weeks after hatching and may then last for several weeks. Another important difference is that in filial imprinting the bird makes its response, that of following, at the same age as it is undergoing imprinting. In sexual imprinting, by contrast, the sexual response is not performed until maturity, many weeks or months after the imprinting has taken place. This can make experiments on sexual imprinting difficult to interpret, because it is not always clear to what extent the animal's experience between the end of the fostering period and the onset of maturity affects its subsequent sexual behaviour. For example, an experimenter is usually faced with the choice, once the fostering period is over, of keeping the young birds either in isolation or housed with other young birds of the same species. The effects of these two situations on the birds' development are likely to be markedly different. Failure to take such influences into consideration may be the reason why different investigators have come to different conclusions about the reversibility of sexual preferences.

The mechanisms controlling the beginning and end of the sensitive period of sexual imprinting are poorly understood. Since the sensitive period may not begin until the bird is some weeks old, it is not possible, as it is in filial imprinting, to suppose that the beginning of the sensitive period depends upon adequate development of the animal's sensory and motor abilities. Precocial birds, especially, are highly mobile and their sensory systems are well developed before the beginning of the sensitive period for sexual imprinting. An obvious suggestion is that the timing of sensitive periods is linked to hormonal changes. However, despite earlier evidence that appeared to support this idea (Immelmann 1972), it now seems that sexual imprinting can take place without the action of endogenous hormones such as testosterone. Thus Hutchison and Bateson (in press) found that sexual imprinting took place in male Japanese quail that had been castrated. As with filial imprinting, the bird's experience during the sensitive period is likely to hasten the end of that period. As one object becomes familiar, so other objects which are novel will tend increasingly to be avoided.

As with filial imprinting, some stimuli are more effective at eliciting sexual imprinting than others. In general, most species imprint sexually most easily on their own species rather than on others. Thus a male zebra finch which is raised by a mixed pair of zebra and Bengalese finches will imprint sexually on the zebra finch regardless of whether the parent zebra finch is the male or the female. It could be objected that parents are more likely to respond to young of their own species than to the young of others, and so the young is simply imprinting on the more responsive parent. Recent research on this question has produced conflicting results, and so at present it is uncertain to what extent parental attention and to what extent the species characteristics of the parent respectively contribute to sexual imprinting.

The functional questions that can be asked about sexual imprinting are intriguing. Why should animals learn so early in life the characteristics of individuals for which they will later show sexual preference? Why should learning be needed at all? Why should the sensitive period be later than that for filial learning? Why should there be sex and species differences in the nature of the imprinting? Learnt rather than unlearnt responsiveness would presumably be desirable if it is important to the young animal that it responds differently to individuals who differ only slightly from each other in appearance. A suggestion put forward by Bateson (1979) is that it is advantageous for young birds to breed with conspecifics that are slightly genetically dissimilar from themselves, but not too much so. If one assumes that the extent to which two animals are genetically similar is detectable in some way by the extent to which they resemble each other, then it would be advantageous for an animal to breed with individuals that diverged slightly, but not too much, in appearance from that animal's immediate relatives. Bateson argues, therefore, that young birds need to learn the individual characteristics of their parents and siblings, so that they can later choose to mate with individuals that diverge slightly from the characteristics that they have learnt. This learning would presumably need to take place early in life because it is then that the young animal is continually close to its relatives over a long period of time. Bateson argues that sexual imprinting occurs later in life than filial imprinting, because it is only later that the young bird's siblings start to fledge and acquire the characteristics that they will retain into adulthood. Certainly the 'optimal discrepancy' hypothesis, as it can be called, makes the interesting and testable prediction that birds will prefer to court individuals that differ slightly but not too much in appearance from an individual with which they were raised, and this has indeed been found to be the case in Japanese quail (Bateson 1978b). Moreover, Bateson (1982) has shown more recently

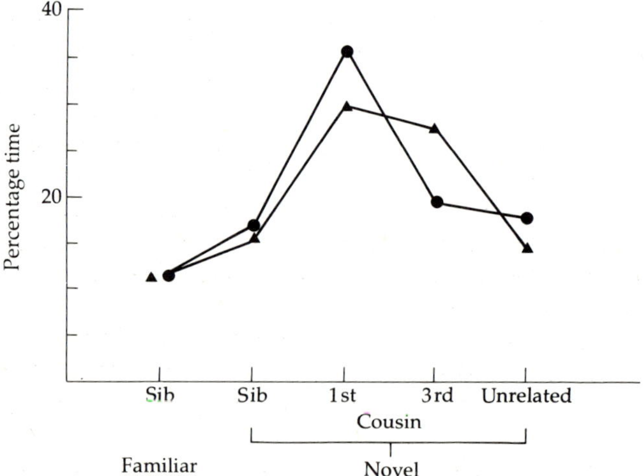

Fig. 4.4. The mean percentage of time spent by adult Japanese quail near members of the opposite sex that were either familiar siblings (sib), novel siblings, novel first cousins, novel third cousins, or novel unrelated individuals. The chance level is 20%. ▲ males (n = 22), ● females (n = 13). (From Bateson 1982.)

that quail that have been reared with their siblings subsequently prefer a first cousin of the opposite sex to more or less closely related individuals (Fig. 4.4). Nonetheless, it is still puzzling as to why the young bird needs to wait until its siblings fledge until it learns the appearance of its kin. Why can it not simply learn the appearance of its parents? What exactly *does* it learn from its parents, and what from its siblings? Indeed, in sexual imprinting, as in filial imprinting, the respective roles of the parents and the siblings in developing the young bird's preferences are far from clear, and are in need of further investigation.

There is a need to find functional explanations, not only for sexual imprinting *per se*, but also for the variation in its characteristics that is found between the sexes and among species. In many ducks and geese, for example, males show sexual imprinting whereas females do not. In estrildine finches such as the zebra finch, sexual imprinting occurs rapidly, early in life, and is virtually irreversible. In other species, such as the bullfinch, it occurs later in life, more slowly and remains reversible for a long period.

Sex differences within a species may occur for a variety of reasons. Several authors have suggested that in those species where the adult female alone cares for the young, sexual imprinting should not be shown by young females because this would lead later in life to attempted homosexual pairings. In species where both parents care for the young, and where the males are more conspicuous in their appearance or behaviour than the females, it has been suggested that the task of mate recognition is more difficult for young males than for young females. Males therefore need to learn the characteristics of mates of their species whereas females can successfully rely on less precise, non-learnt mechanisms of mate recognition.

Species differences in sexual imprinting may reflect differences in their natural history. Estrildine finches frequently breed in colonies containing several species, and they form pairs rapidly as soon as environmental conditions are appropriate. Their early, rapid and irreversible sexual imprinting may be an adaptation to this situation. The young birds undergo sexual imprinting before they are active enough to move about and meet birds of other estrildine species nesting nearby. By contrast, bullfinches have solitary nests and form pairs slowly, so their young do not need to undergo rapid and irreversible imprinting early in life. As with all speculations about the adaptive significance of behaviour, a great deal of further evidence across numerous species is needed before these suggestions become truly convincing, but they are valuable in that they suggest the kinds of information that field biologists should be seeking.

Long-term effects of the quality of the mother–infant relationship

Sexual imprinting illustrates clearly that the early social experience of an animal can profoundly influence its subsequent development. There is now a considerable amount of evidence that quite subtle features of the mother–infant relationship can have effects on the course of the developing animal's behaviour, some of which are very long lasting. Much of this evidence comes from studies of primates. For example, Hinde (1977) separated infant rhesus monkeys from their mothers for six days when the infants were between 21 and 32 weeks old. The separation was achieved by removing the mother from the social group in which she had been living, but leaving the infant behind in the group. The infant's behaviour was recorded during the period of separation, and both the mother's and the infant's behaviour were recorded before the separation and after their reunion. Hinde found that the infants were greatly upset by the separation; they gave frequent distress calls and tended to remain immobile for long periods of time. Hinde also found that the infants most distressed in the days following the reunion were those that had been rejected most by their mothers before the separation and that had played the biggest part in maintaining proximity with the mother.

Further evidence for long-lasting effects of the mother–infant relationship comes from a study by Dunn (1976). She investigated differences in the relationships of human mothers and their babies at several ages from the first ten days of the baby's life through to four years, and analysed to what extent differences that appeared early in the baby's life correlated with differences that appeared later on. Several significant positive correlations emerged. For example, mothers who talked most to their babies at 14 months and who responded to the highest percentage of the baby's utterances were those who, in the first ten days of the baby's life, had most frequently smiled and looked at and talked affectionately to their babies, and whose babies had sucked at the highest rate during a sucking test carried out when they were 8 days old. Moreover, the babies of such mothers had the highest percentage of demands in their utterances at 14 months. Also, children whose mothers' conversation at 14 months tended to be accepting rather than rejecting in character did better at four and a half years on a Stanford Binet IQ test.

Dunn has emphasised an important point which applies not only to her own study but also to other studies which are based on correlational evidence of this sort. She points out that it is impossible to attribute differences in the mother–infant relationship that emerge as the infant grows older exclusively to the mother or exclusively to the infant. The mothers who talk affectionately the most to their babies in the first ten days after birth have babies with the highest sucking rate at that age. Possibly, the mothers talk affectionately the most because their babies suck at the highest rate, or vice versa. Possibly, the babies' sucking rate, even as early as the second week of life, is influenced by the mother's behaviour to it during the first week or, again, vice versa. Possibly, the mothers who talked most to their babies both in the first ten days and at fourteen months were simply garrulous by nature but, equally possibly, early talkativeness may induce changes in the baby, which in turn produce further changes in the mother as the months pass. Possibly also, both the mother's talkativeness and the baby's responsiveness are affected by circumstances external to both of them, such as the behaviour of other members of the family. In summary, in an intimate and long-lasting relationship, such as that between a mother and her infant, it can be extremely difficult to separate the respective contributions of each party to that relationship.

4.2.4 *Disturbances in early development*

The care which a mother normally offers her infant is open to disruption from many sources: the mother may die, be injured or taken ill, or she may be absent for lengthy periods. It is important to discover what effect such disruption has on the infant's development. In human beings it is particularly important to discover whether the disruption of an infant's relationship with its mother, or with whoever acts as its principal caregiver, has long-lasting and deleterious effects.

There are two main methods of attacking this problem. The first is to disrupt experimentally one or more of the infant's relationships under controlled conditions, and observe the effects on development. This can only be done with non-human animals. The second is to survey the development of numerous individuals, taking note of the kinds of relationships they have with various other individuals early in life, and of any disturbances of these relationships. By surveying enough individuals, it is then possible to see whether any of these features of early life predictably leads to given behavioural traits later in life.

Perhaps the best-known experiments involving the disruption of infants' relationships with other animals were carried out on rhesus monkeys by the Harlows and their colleagues. (These experiments are summarised by Mineka and Suomi (1978).) Baby captive rhesus monkeys that were kept in total isolation from their mothers and from other animals for several months showed gross behavioural abnormalities both during the period of isolation and after they had been returned to the company of other rhesus monkeys. During the period of isolation, they initially gave frequent distress vocalisations and later subsided into periods of immobility in which they sat in a hunched position sucking their fingers or toes. These two phases, which prove to be characteristic of such periods of separation, have subsequently been called periods of 'protest' and 'despair'. When they were returned to their mothers and to the company of other rhesus monkeys, the formerly isolated monkeys (hereafter termed 'isolates') showed severe and long-lasting behavioural disturbances. They were involved in frequent fights with other group members. They often seemed to provoke these by staring directly at their companions (a rhesus threat gesture) and they also threw themselves into attack against high-ranking members of their group. Normal monkeys would have avoided both of these behaviours. The isolates tended to bite themselves, they developed stereotyped movements such as rocking and, when they matured, they showed abnormal sexual behaviour. Despite these abnormalities, some of the isolated females became pregnant, thanks to the activities of extremely persistent males, but when they gave birth they proved to be very abnormal in their maternal behaviour, often causing severe physical harm to their infants.

A great deal of research has been carried out in order to discover the range of effects produced by isolation, and also to discover how these isolation effects vary with the nature of the isolation procedure and the animals used in the experiments. Most, but not all, of these experiments have been carried out on rhesus monkeys. The conditions under which the animals are kept when isolated greatly affect the nature of their response. Not surprisingly, the effects are most severe when the animal is totally isolated visually and aurally from other animals and confined in a small cage. The effects are less severe if the conditions of isolation are relaxed: infants kept in separate cages from other animals, but which can see and hear them, are less severely disturbed than those which are totally isolated. Long periods of isolation prove to be more damaging than short periods, though even isolation for a period as short as 13 days has long-lasting effects (Hinde 1977). There are certain ages, around three to nine months in the rhesus monkey, when the infant is especially sensitive to isolation. Males prove to be more upset by isolation than females.

If a baby rhesus monkey is isolated from all other animals, it is of course impossible to say whether the effects of the isolation stem from the absence of the mother, the absence of other companions or both. To resolve this issue, Harlow and his colleagues raised rhesus babies either with their mothers but without other infants, or vice versa. The results were conflicting in some respects, but showed that babies reared under either condition displayed behavioural abnormalities compared with babies reared normally. These abnormalities were not as severe, however, as in animals which had been kept in total isolation. Nevertheless, the experiments show that both the mother and other infants are necessary for normal behavioural development.

The Harlows' first experiments on isolated rhesus monkeys provoked widespread interest, particularly among child psychiatrists, such as Bowlby, who were concerned with the effects of separation on child development. Bowlby (1951) had gathered evidence suggesting that children who were prevented from forming a lasting relationship either with their mother or with a mother-substitute, or children whose relationship with their mother was severely disrupted for a long period of time, were more likely than others to show conduct disorders later in life. They were more likely, for example, to appear before juvenile courts for repeated delinquency, and more likely to be removed from home and taken into care by their local authorities. They were also more likely to form only shallow ties of affection with other people.

It was difficult to be certain from Bowlby's evidence that disruption of the infant's relationship with its mother was the prime cause of these problems. A child raised in an institution, for example, might become a persistent offender for many reasons other than having no permanent relationship with its mother or a mother-substitute. Nevertheless, Bowlby's work alerted people to the possible dangers of disrupting a relationship between a mother and her baby, and stimulated a lot of further research, much of it consisting of surveys of the kind mentioned earlier. The results of these surveys have been summarised by Rutter (1979). Rutter concludes that the main factor that operates early in life to promote conduct disorders is family discord and disharmony, rather than separation itself. Thus parental divorce early in the infant's life was

Fig. 4.5. A mother yellow baboon with her newborn infant, surrounded by several juveniles. (From Altmann 1980.)

strongly linked with later delinquency, whereas parental death was not. Moreover, children who remained in homes where there was continual marital discord showed greater behavioural disturbance than children whose parents had divorced, and whose life had thereafter become more tranquil.

There is now a considerable amount of evidence, both from the surveys of children already mentioned and from experiments by Harlow, Hinde, Mason and others, that the role of the mother is not simply to fulfil the infant's physical needs. Rather, it suggests that, as the mother–infant relationship endures, a strong bond builds up between the two individuals, and it is this bond that is essential for normal development. The formation of the bond can be impeded either by separating the mother and infant or by otherwise denying mother and infant the opportunity to attend to each other within a harmonious social environment.

4.3 Infant relationships with animals other than the mother

4.3.1 Supportive relationships during competition

Although an infant's relationship with its mother is, in many species, undeniably important, infants may also develop close relationships with other members of their social group, such as their father, aunts, brothers and sisters. Which of these relatives are available to the infant, and how many of them are available, varies from species to species. In most song-birds, for example, the young bird's only social companions are its fellow nestlings, its mother and its father. Older brothers and sisters from previous broods have already dispersed. In many mammalian and some avian species, however, adolescent females, or males, or both, stay with their parents and can therefore provide potential companionship for their younger siblings (Fig 4.5). (These may be full or half brothers and sisters, depending upon whether they have the same father or different fathers.) Sometimes, the newly matured animals may remain in the group for only a limited time before leaving, but in many species they may spend their entire lives in the group into which they were born. This happens, for example, in many ungulates and primates and some carnivores, and in the great majority of cases it is the female who stays in her natal group, whereas the male leaves at maturity. In such societies, referred to as matrilineal, the young female can develop long-term relationships with numerous related females, such as aunts and mature sisters.

The relationships that develop between the infants and other members of matrilineal societies can be extraordinarily subtle and complicated. The animals in which these complexities have been most fully worked out are Japanese macaques (*Macaca fuscata*) and rhesus macaques. In rhesus monkeys, for example, Berman (1981) and others have shown that if an infant or juvenile becomes involved in a dispute it is likely to receive help from close female relatives, i.e. its mother, aunts and older sisters. Moreover, infants that have high-ranking close female relatives are more likely to receive support than infants with low-ranking ones. Infants with high-ranking close female relatives almost invariably attain a high rank themselves as they grow up. This suggests strongly that the rank that an infant acquires is determined by its relatives' behaviour towards those with which it interacts. An animal will submit to an infant in a dispute if the infant is constantly supported by high-ranking relatives, and will not submit to it if its own relatives offer stronger support than the infant's. Thus an infant will gradually learn its own rank from the responses of those with which it interacts. Rhesus and Japanese macaques therefore provide an example where an individual's position in its society, which includes its ability to gain access to vital resources, is shaped by the nature of the relationships that it develops with its relatives as it matures.

4.3.2 *Infant care by non-mothers*

Infants commonly receive care from animals which are not their mothers. Paternal care is widespread in the animal kingdom (see review by Ridley (1978)), and in many species the father offers very similar care to that given by the mother. For example, in many birds both parents share equally in incubating the eggs and feeding the nestlings and, among mammals, mother and father Mongolian gerbils (*Meriones unguiculatus*) possess similar repertoires of care-giving behaviour, although the males cannot of course lactate. The two animals build a common nest and spend well over half their time in it during the first two weeks of their pups' life. Also, both lick and sniff their pups and spend a high proportion of their time in contact with them (Elwood 1975). Young animals clearly benefit from paternal care. For instance, all except the smallest and least viable pups of the California mouse (*Peromyscus californicus*) had a higher probability of surviving in experimental litters where the father was present than in litters where he was absent (Dudley 1974). They also developed faster. Dudley attributes their better survival and faster development to the fact that the father helped to keep them warm.

Although the father can directly affect an infant's development, he can also do so indirectly by influencing the mother. For example, the male Mongolian gerbil spends some of his time in the nest with his pups, and the mother spends correspondingly less time in the nest than if the male is not present in the cage with her. The mother's workload in caring for her pups seems to be reduced by the father's presence, and there is evidence that she is less stressed as a consequence (Elwood & Broom 1978). It is possible, therefore, that the mother's more relaxed behaviour in turn affects the pups' development. To complicate matters still further, the father's presence may directly affect the pups, and their changed behaviour may in turn influence the mother. The father helps to keep the pups warm, thereby reducing their distress calling, and also shields the mother from their olfactory stimuli. This in turn reduces the amount of nest building that mothers perform, and the amount that they sniff the pups. These changes in the mother's behaviour may again affect the pups' development. Even in a small family group of gerbils, therefore, where there is a relatively simple network of relationships, the route by which paternal effects are exerted may be surprisingly complicated.

Among the marmosets and tamarins, which are South American monkeys that live in small families, both the father and the older brothers and sisters spend a great deal of time carrying the infants. Indeed, in the common marmoset (*Callithrix jacchus*), fathers may carry the infants more than the mothers do during the early weeks of infant life (Locke-Haydon & Chalmers 1983). As the infant marmoset matures, it is carried less by its relatives; it is possible to assess the responsibility of each of the animals for the change in their relationship, just as it is possible in the simpler case of the rhesus monkey described in section 4.2. Again, the change is not simply due to the infant. The mother, father and older siblings all play a part in promoting the infant's independence by rejecting it more often as it grows older, and by initiating fewer episodes of carrying.

Marmosets and tamarins are just one among many groups of animals where individuals other than the parents care for the infant. 'Allomaternal behaviour', as it is frequently called, is found in many mammalian groups, ranging from insectivores and rodents to carnivores and dolphins, and it is also found in some birds. In the Florida scrub jay (*Aphelocoma coerulescens*), year-old birds hatched during the previous breeding season help by feeding the young, defending the territory from other jays and detecting predators (Woolfenden 1975).

The causal factors that promote paternal and allomaternal care are not well understood. Certainly, the nature of the stimuli emanating from the young animal is important, just as it is in the case of maternal behaviour. Virgin female mice will respond more strongly to living than dead one-day-old pups, just as will mothers (Noirot 1964, 1965). Altmann (1980) has produced evidence suggesting that baboon infants become less attractive to other group members between four and six months when they lose their distinctive black natal coat and pink skin, and start to acquire adult colouration.

The factors controlling the onset of rodent paternal behaviour are particularly interesting. The adult males of laboratory mice and Mongolian gerbils, amongst other species, will under certain conditions kill and eat their own pups, and yet under other conditions they will show paternal behaviour. Some factor or factors must therefore suppress the male's cannibalistic tendencies and promote infant care. Elwood (1980) has identified some of these factors in the Mongolian gerbil. A male which has never seen newborn gerbil pups, and which is housed with a female before she becomes pregnant, will kill and try to eat any pups that are introduced into his cage. Once the female becomes pregnant, however, the male shows a decreasing tendency to kill any pups put into the cage, and an increasing tendency to care for them as his mate's pregnancy proceeds. The female's pregnancy only confers temporary immunity upon the pups, however. Following removal of the pregnant female, the male becomes increasingly likely to kill any pups put into the cage. If, however, the female stays with the male until she gives birth, the male will then help to look after the litter. This experience irreversibly suppresses the father's cannibalistic behaviour.

From an adaptive point of view, it is interesting to ask what benefit, if any, individuals derive from helping to rear the offspring of others. First, the offspring themselves may have a higher probability of surviving if they receive this extra help than if they do not. Woolfenden showed this to be the case in Florida scrub jays. Moreover, the helpers were in most instances older siblings of the nestlings which they were helping. Consequently, a certain proportion of the nestlings' and helpers' genes were identical by virtue of their descent from common parents. By promoting the survival of their younger siblings, the helpers were thus encouraging the spread of identical copies of genes that they themselves carried. To use current terminology, they were increasing their inclusive fitness. Secondly, by helping younger siblings, older siblings may improve their own skill at caring for infants. This seems to be the case in the tamarins and marmosets. There are several reports of captive females, who have been denied the opportunity to care for younger siblings, giving inadequate care to their own firstborn young; they may even kill them. By contrast, females who have looked after younger siblings usually care for their own first offspring without difficulty.

When adolescents share in caring for their young siblings, therefore, they may increase their individual fitness by improving their skill as a caregiver, and increase their inclusive fitness by increasing the viability of their younger siblings.

4.3.3 Play

One of the most striking and yet enigmatic behaviours that infants perform amongst themselves is social play. It is enigmatic because it is difficult to define, its function is not at all obvious, and yet it is widespread among mammals. (It also occurs in some birds such as parrots, rooks and crows.) One reason why play is difficult to define may be that it is probably in reality a heterogeneous collection of behaviours that possess little in common either in terms of the mechanisms that control their performance or in their consequences for the young animal's development. Be that as it may, there are certain components of playful behaviour which can easily and reliably be identified and about which there is little disagreement as to their playful nature. Many mammals, including carnivores and primates, indulge in bouts of mouthing and wrestling (Fig. 4.6), distinguishable from true fighting by the relaxed posture of the participants and by the absence of physical violence. Many also indulge in playful chasing, which is again relaxed in character and often involves the chaser and chased animals frequently switching roles.

Social play poses numerous questions for the biologist. How does it develop? What causes it to happen? What use is it? Why has it evolved? Several studies have shown that play develops in a predictable sequence, reaching a peak before adolescence and then dying away. In baboons, for example, wrestling play reaches a peak in two and three year olds (Chalmers 1980b). The different components of play develop at different rates. Typically, wrestling play develops earlier than chasing. Chalmers (1980a) has shown how wrestling in wild baboons develops from the young animal's earliest abilities to hold and mouth objects, and how the animal gradually distinguishes between inanimate objects that can be eaten, inanimate objects that can only be manipulated, and fellow

Fig. 4.6. Rough-and-tumble play beween an infant and a juvenile marmoset, *Callithrix jacchus*. (From a photograph by Miranda Stevenson.)

baboons that can be wrestled with. Owens (1975), also working on wild baboons, has shown how wrestling play becomes more like true fighting as the animals grow older. The relationship between play and 'serious' behaviour can also be seen clearly in predatory mammals, such as cats and dogs. Here, behaviours that are used in hunting, for instance stalking, pouncing and biting, are prominent components of social play.

The factors controlling the performance of play are poorly understood. Play is known to be affected by hormones. Young males of many species typically play differently from young females, and genetically female rhesus monkeys whose mothers are injected during gestation with the hormone testosterone play later in life in a way that is intermediate between that of normal males and normal females (Goy & Phoenix 1971). However, play is also influenced by the availability and behaviour of potential playmates. Particularly striking is evidence from cats which shows that a female kitten is more likely to play with objects if there is a male in her litter, than if all her litter-mates are females (Barrett & Bateson 1978). Finally, play in kittens can also be influenced by the advent of weaning. The amount of time that kittens spend playing with objects rises to a peak just after weaning is complete, and social play with other kittens peaks a little earlier. However, if weaning is advanced artificially, the peaks for both object and social play become higher (Bateson & Young 1981). Early weaning reduces the time available to the kittens for play, since after weaning the young cats would, under natural conditions, leave the den and disperse. Bateson and Young therefore suggest that when weaning is advanced artificially the litters indulge in extra play, thereby compensating for the reduced amount of time that is available to them.

The function or functions of play are as poorly understood as its causation. Two main theories are, first, that play develops the young animal's physical abilities—its muscular development, co-ordination, lungs and heart— and, secondly, that play develops its social skills. Young animals are presumed to learn the behavioural characteristics of their group members, and the niceties of social communication, in the non-serious context of play rather than in potentially dangerous serious competitive encounters.

The evidence for these theories has been reviewed in a number of publications (e.g. Fagen 1981; Smith 1982). Some authors have argued that the kind of play that young animals perform develops skills that are peculiarly suited to their later way of life. Byers (1980), for example, found that young Siberian ibex (*Capra ibex sibirica*) used the same motor patterns in play, such as mounting, pushing and butting, as older animals used in contests and when mating. Moreover, they selectively played with young animals of their own size. Larger male kids in their first year of life played frequently with each other, but not with smaller females of the same age or with smaller and younger males. They also played frequently with females a year older than themselves, which were only slightly larger, but they did not play with males a year older than themselves, which were much larger. Since a kid cannot easily mount, push or butt another kid which is much bigger or smaller than itself, Byers argues that kids choose those play partners which offer opportunities to play most vigorously.

Direct experimental evidence for the developmental consequences of play is, however, hard to find. This is because it is difficult either to deprive an animal of the opportunity to play, or to enhance its opportunity to play, without affecting many other aspects of its behaviour as well. For example, some early experiments by Alexander and Harlow (1965) on rhesus monkeys showed that infants that were raised with their mothers but with no playmates developed several behavioural abnormalities. It is impossible to say from this, however, whether it is the absence of play or the absence of some other feature of the age-mates' presence that led to the abnormalities. Recently, more decisive experimental evidence has been produced by Einon, Morgan and Kibbler (1978). They showed that the quality of the interactions in which juvenile rats engaged affected their behaviour later in life. The animals were housed between 25 and 45 days of age in different conditions. Some were kept in small groups, some in isolation, and others were isolated except for a period of one hour a day when each was housed with another rat of the same age, which was also kept in isolation at other times. Finally, some rats were kept in isolation and were allowed one hour of contact per day with another rat of the same age, but this second rat was drugged. Some of these drugged rats received chlorpromazine, and were almost entirely passive throughout the hour-long sessions; others received amphetamine and spent much of their time sniffing and following their partner. By contrast, when undrugged animals were housed together for an hour after a period of isolation, they spent much of the time playfully chasing and wrestling. All of the animals were then subjected to a series of tests starting when they were 45 days old. One test involved placing each rat in an unfamiliar arena (an 'open field'), and measuring how quickly the rat ceased exploring the arena and any strange objects placed in it. The rats that had been reared in small social groups rapidly ceased their exploration, whereas those that had been reared in total isolation continued to explore for a long time. Rats that had been allowed one hour of social contact per day with an undrugged partner were intermediate between the other two groups but more like the socially raised than the isolated rats. Rats that had been housed with a drugged partner for an hour a day, however, were more like animals that had been raised in complete isolation than any others. Further tests revealed a similar pattern among the different groups of rats. In particular, socially reared rats were best able, and totally isolated rats least able, to reverse a previously learned habit, with partially isolated animals intermediate in their abilities.

Although these results strongly suggest that playful interactions between 25 and 45 days with age-mates are important for the development of exploration and learning, they are still not totally conclusive. The young rats did not play with their drugged partners, but neither did they engage in other normal social behaviours and it may be that the absence of these other behaviours contributed to some of the later behavioural abnormalities.

This brief account of social play shows that it is still a poorly understood aspect of behaviour. Even if it becomes clearly established that social play does affect the development of physical and social skills, this is only the first of many issues that have to be resolved. Both social play and social skills have many components, and it is important to discover which aspects of social play promote which skills and, above all, the mechanisms by which they achieve this effect.

4.4 Conclusions

It is difficult to draw together the rather disparate collection of phenomena and investigations that come under the heading of the development of social behaviour. Certainly, the material covered in this chapter is richer in descriptive detail and *ad hoc* theories than in general principles. Given the present state of knowledge of developmental phenomena, one can do no more than guess which areas of study are most likely in the future to reveal generally applicable principles of development. From the account given in this chapter, three topics suggest themselves as candidates. First, although there is certainly no single mechanism that determines the duration of sensitive periods or the stage in development at which they occur, further studies of the way in which animals' responsiveness to social objects changes as they develop may reveal a group of mechanisms at work, among which certain patterns and regularities may become clear. Secondly, given that many of the behaviours of both partners in a relationship affect, and are affected by, each other, it may become possible to discover general rules governing the behaviours that are particularly influenced in this way and other general rules governing behaviours that are particularly resistant to the partner's behaviour. Finally, given that much of social development is flexible, for evolutionary reasons that have been briefly described in this chapter and are more fully discussed by Fagen (1981), a picture may emerge of how animals so often manage to develop into socially competent adults with the repertoire of social behaviour typical of their species, despite being exposed during infancy to many environmental influences that are liable to blow them off course.

4.5 Selected reading

Hinde (1979) develops the concept of the social relationship, particularly with respect to human beings. Chalmers (1979) reviews much of the literature on the development of mother–infant relationships in human and non-human primates. A valuable summary of the effects of separation, family disruption and other forms of disturbance on human infant development is given by Rutter (1979). Dawkins (1976, 1982) provides a thorough discussion of parent–offspring conflict, in the context of a much broader discussion of natural selection and behaviour. The extensive literature on imprinting can usefully be approached by reading Bateson (1979). The literature on the nature and functions of play is reviewed critically by Smith (1982), with the bonus that Smith's paper is accompanied by critiques written by several others currently studying play. Ridley (1978) reviews paternal behaviour.

CHAPTER 5
GENERAL PRINCIPLES OF LEARNING

N.J. MACKINTOSH

5.1 Introduction: definition of learning

The behaviour of an animal is often fixed and unvarying, in the sense that the same stimulus always elicits much the same response. Dogs flex their foreleg whenever their paw is pricked; rabbits blink whenever a puff of air is directed towards their eyes; the female greylag goose always retrieves an egg that has rolled out of her nest. Most such responses have obvious adaptive significance, and if a particular response is indeed always appropriate in a particular situation it is not surprising that it is always elicited, with some fixed connection between stimulus and response. But it is equally obvious that a response that was once appropriate in a given situation may now no longer be so. Circumstances may change and behaviour, if it is to remain adaptive, must change with them. Yesterday, a foraging animal found food in a particular patch; but if it continues to search there today, when the food has all been consumed, it may starve to death.

When we observe that an animal's behaviour changes in response to some change in its environment, we may attribute this change to learning. Just as the fixed behaviour patterns of a species may change over many generations in response to slow environmental changes, so individually learned patterns of behaviour may change in response to abrupt or transient environmental changes. Although this analogy between the natural selection of advantageous patterns of behaviour and the modification of learned behaviour should not be pressed too far, it is helpful if it reminds us that the ability to learn would not have evolved had it not been of some adaptive significance, and that particular instances of learning can therefore be expected to have advantageous consequences for the learner.

We are all familiar with the phenomenon of learning but, as is often the way with a familiar phenomenon, it is not easy to provide a precise, scientifically useful definition of it. Experimental psychologists have sometimes identified learning with observable changes in behaviour, as when Kimble (1961), in a widely used textbook, wrote that 'learning refers to a more or less permanent change in behaviour which occurs as a result of practice'. But this definition does not accord with any usual meaning of the term. In common usage, we *infer* (or may infer) that an animal has learned something when we observe a particular change in its behaviour. Learning is an inference from, or explanation of, a change in behaviour. The first problem, then, is to distinguish between those changes which we wish to attribute to learning and those which should be attributed to other causes. If an animal starts searching for food in a particular location, although it did not do so six hours ago given an equal opportunity, this may not be because it has learned anything in the interval. It is equally possible that it is hungrier now than it then was; its state of motivation may have changed. Some longer term changes are usually attributed to an ill-defined process of maturation. A male puppy urinates in the same way as a female, by squatting, but an adult dog cocks his hind leg. The change is not a consequence of learning, but of increasing sexual maturity; if the puppy is injected with male hormones it will promptly behave like an adult.

In these and other cases, we are confident that the change in behaviour cannot be due to learning, because the animal has had no opportunity to learn. This suggests that learning depends on certain definable types of experience, as in Kimble's definition where learning is said to depend on practice. But even if we know what we mean by 'practice', a further qualification is necessary. A change in behaviour that occurs without the opportunity for practice or other relevant experience may confidently be ascribed to causes other than learning, but the converse implication does not hold. An animal may appear to engage in practice and its behaviour may subsequently change, but experimental analysis is required to prove that the practice was necessary for the change in behaviour. Young birds, for example, cannot fly. As they grow older, they can be seen practising, and after a few weeks they can fly quite proficiently. It seems reasonable to suppose that they must learn to fly and that the period of practice is necessary for the development of this skill. But pigeons have been brought up under restricted conditions which completely prevented their practising flight movements. When they were released at the age at which normally reared birds were first flying successfully, there was essentially no difference in the proficiency of those that had practised and those that had not (Grohmann 1939).

Thus, learning may be defined by reference to the particular set of circumstances that is responsible for the observed changes in behaviour. If motivational changes are those produced by one set of conditions (e.g. by depriving an animal of food or water), then changes attributable to learning are those produced by circumstances which provide certain other experiences. The definition will be of limited value, however, unless those circumstances can be specified. It is not very much use to say that learning is produced by practice, unless the concept of practice is further analysed.

Further analysis is undoubtedly possible in particular cases. Sufficiently precise definitions can be provided of the circumstances responsible for specific kinds of learning and the change in behaviour such learning produces. A useful definition of classical conditioning, for example, would refer to particular sorts of changes in an animal's behaviour to a particular stimulus, brought about by exposing the animal to a correlation between that stimulus and another. However, psychologists and ethologists have devised a large number of different procedures for studying learning in animals, and the variety is such that it may be difficult or impossible to formulate a definition of the conditions producing learning that is neither so vacuous that it says nothing and fails to exclude cases which do not seem to constitute learning, nor so restrictive that it excludes certain cases which we should certainly want to regard as instances of learning.

It is probably foolish to spend too much time attempting to provide a precise, all-embracing definition of learning. A more useful strategy will be to provide precise descriptions of some of the particular situations in which learning has been studied. These definitions will specify the sorts of changes in behaviour that an experimenter may observe, and the conditions or experimental operations responsible for those changes.

A possible consequence of this procedure is that it may lead to fragmentation of the field of learning. A series of distinct definitions of different procedures and different consequences of those procedures may encourage the belief that they have nothing in common with one another, that the term learning refers to multiple unrelated processes rather than to one. This may not, of course, be a bad thing. It is certainly not obvious that the same processes are involved in learning to read music, play tennis and solve differential equations. But in the end, the question is one of theory. Learning is a single process if we can propose a satisfactory theory of learning that will explain the variety of changes in behaviour we observe resulting from the variety of experimental operations we perform. The fact that different operations produce different changes in behaviour does not rule out the possibility that the mechanisms of learning involved are the same. The operations that produce habituation, for example, are quite different from those that produce classical conditioning, and the change in the animal's behaviour observed in the two types of experiment is quite different, but several theories of habituation have attributed it to the same process as that underlying conditioning.

With this qualification in mind, we shall in what follows distinguish between non-associative learning (habituation and sensitisation), simple associative learning (classical and instrumental conditioning) and a third, portmanteau category termed, in ignorance, complex learning.

5.2 Non-associative learning

The distinction between non-associative and associative learning is solely in terms of the experimenter's operations. In experiments on associative learning, animals are exposed to more than one event, standing in a specified relationship to one another, and a consequent change in behaviour is observed. Non-associative learning is being studied when the animal is repeatedly exposed to only a single event. Two consequences of such exposure have been documented—habituation and sensitisation.

5.2.1 Habituation

If a snail is moving along a wooden platform, it will immediately withdraw into its shell if one taps on the wooden surface. After a pause, it will emerge and continue on its way, but will again withdraw if one taps again. This time, however, it will re-emerge more rapidly, and a third and fourth tap may elicit only a brief and perfunctory withdrawal response. After a few trials, the stimulus which initially elicited an immediate response may have no detectable effect on the animal at all. It is said to have habituated (Humphrey 1933).

What is the cause of this change in behaviour? Two rather uninteresting possibilities need to be ruled out before we should want to infer that it is due to a process of learning or habituation. First, sensory adaptation may reduce the animal's sensitivity to a repeatedly presented stimulus; secondly, muscular fatigue may reduce the probability of the response occurring.

Experiments with the sea snail *Aplysia* show how these possibilities may be discounted (Kandel 1976). Tactile stimulation of the siphon or mantle shelf of the animal elicits a gill withdrawal reflex, but this response rapidly declines with repeated presentation of the stimulus (Fig. 5.1). At this point, however, a relatively

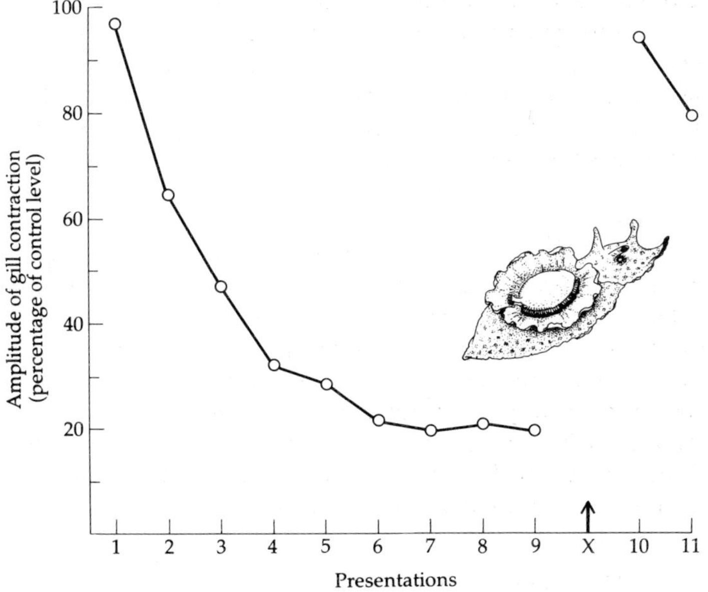

Fig. 5.1. Habituation of the gill withdrawal reflex in *Aplysia* to repeated tactile stimulation. At X, a novel, dishabituating stimulus is applied, and the next presentation of the original tactile stimulus elicits a strong response. (After Kandel 1976.)

strong, novel tactile stimulus may be applied elsewhere and a further presentation of the original stimulus now elicits a gill withdrawal reflex comparable in magnitude to that observed at the outset of the experiment. This phenomenon is usually termed dishabituation, and it suggests that the original decline in responding cannot have been due to either sensory adaptation or muscular fatigue.

Neurophysiological studies undertaken by Kandel and his colleagues have abundantly confirmed this conclusion. Recording from afferent nerves during the course of repetitive stimulation reveals no change in the level of activity, and the gill withdrawal reflex habituates even if it is elicited by direct electrical stimulation of the afferent nerve, bypassing the receptors altogether. Effector fatigue can be ruled out by showing that repetitive stimulation of the motoneurons controlling gill withdrawal produces no decline in overt responding and that direct stimulation of the efferent nerve or motoneurons after the gill withdrawal response has habituated elicits a perfectly normal response. The neuronal changes underlying the habituation of the withdrawal reflex have been identified as being more central. Specifically, repetitive tactile stimulation of the siphon, sufficient to produce behavioural habituation, causes a decline in the excitatory postsynaptic potentials in the motoneurons innervating the gill withdrawal response, and these potentials can be restored by presentation of a dishabituating stimulus.

Identification of the neuronal correlates of this simple instance of habituation should not lead one to suppose that there is no need for further behavioural analysis. For a start, the habituation observed in *Aplysia* may be a simpler phenomenon than that observed in mammals. Habituation in *Aplysia* is a relatively short-term phenomenon; after a sufficient interval for recovery, presentation of the original tactile stimulus elicits the gill withdrawal response at full strength and a completely new series of trials is needed to cause its further habituation. Although such spontaneous recovery is a common feature of most habituated responses, in mammals a response that has been recently habituated will not normally show full recovery and will decline again more rapidly than before if the eliciting stimulus is again repeatedly presented. Typical results are shown in Fig. 5.2. The occurrence of spontaneous recovery implies that there is a short-term

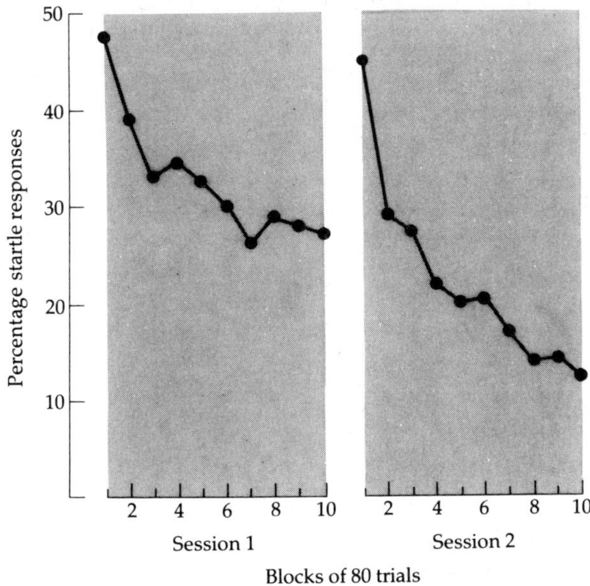

Fig. 5.2. Habituation of the startle response in rats to a repeatedly presented brief noise. Within each session trials occurred at an average interval of 7.5 s. There was a 1-hour interval between sessions 1 and 2. (From Marlin & Miller 1981.)

mechanism of habituation, but the fact that the recovery is incomplete implies that there must also be a longer term mechanism. If habituation depends on establishing some representation of a repeatedly presented stimulus so that it can be classified as familiar rather than novel, then it must be supposed that both transient and more permanent representation can be established. One theory that has stressed both processes has been proposed by Wagner (1976).

Evidence of habituation has been obtained throughout most of the animal kingdom, from coelenterates to human infants. It is particularly noteworthy that habituation has been reliably documented in animals, such as hydra, which have shown no good evidence of being capable of associative learning (Rushforth 1965). This suggests that at least some of the processes responsible for habituation cannot be the same as those required for successful conditioning. On the other hand, it seems relatively certain that the processes involved in habituation cannot be the same in all animals. Indeed, even protozoa may show a decline in responsiveness to a repeated stimulus (Jennings 1906); in this case there can be no question of distinguishing between sensory adaptation, muscular fatigue and central habituation, let alone between short- and long-term mechanisms similar to those presumably operating in mammals.

The generality of habituation implies that it is of considerable biological significance. This is not hard to see. A snail that failed to respond to sudden, novel stimuli by withdrawing into its shell would forfeit much of the advantage of having a shell. But a snail that continued to withdraw into its shell in response to every change in stimulation would have difficulty finding time to eat. Novel stimuli may signify danger and an animal must react to them by appropriate defensive behaviour; but a stimulus which has no further consequences on its first few occurrences will probably continue to be safe, and regularly repeating stimuli are as likely as not to form part of the general background against which more significant events occur. A noise or vibration that occurs repeatedly could well be due to the swaying of a branch in the wind. Thus the function of habituation is to discriminate between novel and familiar events, and to ensure that the animal's behaviour is more or less appropriate to each.

5.2.2 *Sensitisation*

Habituation may be regarded as a mechanism for eliminating unnecessary responses. Learning may also, however, result in the appearance of new responses rather than the disappearance of old. An animal that was formerly quiescent may, as a consequence of a particular set of experiences, start performing a particular response. Most instances of learning in which a new pattern of behaviour is established have been treated by psychologists as instances of associative learning or conditioning. But there is one relatively well-documented case where rather simpler processes may be appealed to.

The common octopus (*Octopus vulgaris*) lives in crevices or openings between rocks, sitting at the entrance to this 'home', emerging to attack crabs or other prey, and retreating inside when danger threatens. Octopuses may be kept in the laboratory in tanks of sea water equipped at one end with a home made of a few bricks, and can then be trained to swim down the tank to attack a food source at the far end or to retreat if such an attack is met with a mild electric shock (Wells 1978). The probability of emerging to attack a neutral stimulus, such as a plastic disc suspended on a rod, is increased if the octopus has recently been given food and decreased if it has recently been shocked. These effects do not depend upon rewarding or punishing the octopus for actually performing the response in question, for they can be obtained by feeding or shocking it in its home. They are not, therefore, cases of conditioning. It seems rather that feeding or shocking the animal increases the probability of the responses of approach or withdrawal normally elicited by food or shock, so that they may temporarily be elicited by a neutral stimulus.

Such an increase in the probability of a response normally elicited by a biologically significant stimulus, consequent upon repeated presentation of that stimulus, is termed sensitisation. It is evident that there must be some conflict between the principles of habituation and sensitisation. In studies of habituation, the repeated presentation of a stimulus results in a decline in the probability of the response elicited by that stimulus. In studies of sensitisation, the same set of operations increases the probability of responding. How can both effects be supposed to occur?

There is no doubt that they do—and in one and the same experiment. What this means is that the decline in responding observed to a repeatedly presented stimulus in an experiment on habituation will be made up of two conflicting processes. It is often possible to observe both in action at once. Thus a common result in habituation experiments is that the animal's responsiveness to the stimulus first increases and is only later followed by a decline (Groves & Thompson 1970). Thus one distinction between the two processes is probably that sensitisation occurs only for the first few presentations of a stimulus, while habituation continues as long as the stimulus is presented. A second factor differentiating them is the intensity or strength of the stimulus in question. Sensitisation appears to be an increasing function of the intensity (and perhaps significance) of the stimulus; events such as food or shock are prime examples of stimuli the repeated presentation of which may produce sensitisation. Weak stimuli appear to produce little sensitisation; they habituate rapidly, and without any evidence of an initial, transient increase in responsiveness.

Recognition of the importance of sensitisation, therefore, is necessary for a proper understanding of habituation. It is also relevant to the analysis of associative conditioning for, as the experiment with the octopus showed, repeated presentation of a significant stimulus may increase the animal's responsiveness to the point where a normally inadequate stimulus temporarily elicits the response in question. Such a change in responsiveness may easily be mistaken for conditioning. How would one have interpreted the results if the octopus had been fed or shocked only when it had emerged to attack the neutral stimulus?

An experiment with marine worms (*Nereis diversicolor*) further illustrates the point (Evans 1966a, b). The worms were placed in horizontal tubes and either fed or shocked at intervals. For one conditioning group, the delivery of food was always preceded by the brief illumination of a light; for the second conditioning group, the delivery of shock was always preceded by a brief decrease in illumination. Both groups showed appropriate changes in behaviour to these stimuli, the fed group being progressively more likely to move down the tube in response to the light, and the shocked group being progressively more likely to withdraw. The natural interpretation is that these changes of behaviour were instances of conditioning, reflecting an association between the change in illumination and the delivery of food or shock. Fortunately, two further groups were run in the experiment, one exposed to food, the other to shock. Interspersed between these presentations of food or shock was a series of test trials on which the light was briefly switched on or off. Exactly the same changes in behaviour were observed in these two groups as in the conditioning groups. Although the change in illumination was in no sense a signal for food or shock, it came to elicit approach or withdrawal depending on whether the animal was being repeatedly fed or shocked. The change in behaviour must be attributed to sensitisation, and it follows that sensitisation was almost certainly equally responsible for the change in behaviour observed in the conditioning groups.

Sensitisation can thus produce changes in behaviour which, in the absence of careful experimental analysis, may be mistaken for instances of associative learning or conditioning. This is unfortunate, for it has meant that sensitisation is usually treated simply as an inconvenient nuisance which has to be controlled for in more interesting studies of conditioning. There has, therefore, been very little systematic investigation of the range of conditions under which sensitisation occurs, the sorts of behavioural changes it may produce, and the underlying processes responsible for those changes. It seems likely, however, that it represents a fairly widespread form of learning, at least among invertebrates; it may indeed be a necessary precursor to associative learning or conditioning. Nor is it difficult to see how such a form of learning could be of considerable adaptive significance. Provided there is some regularity in its environment, an animal which shows an increased propensity to leave its home and explore when food has recently been obtained will have a better chance of obtaining more of the same food; conversely, one that shows an increasing tendency to withdraw to its home following recent exposure to severe danger stands a better chance of escaping the lurking predator. The process of sensitisation thus enables an animal to take advantage of any statistical regularity in its environment without requiring it to learn which specific events are correlated with food or danger. However, sensitisation can be regarded as only a precursor of associative learning, for it seems obvious that associative learning must be the prime method by which an animal discovers what precise events signal the occurrence of events of significance such as food or danger.

5.3 Associative learning

5.3.1 Classical and instrumental conditioning

The laboratory study of learning in animals has long been dominated by the study of associative learning or conditioning. In a typical conditioning experiment, the experimenter arranges a temporal relationship between two events, that is to say he arranges that whenever one event (E1) occurs, it is followed, more or less immediately, by the other (E2). He then observes whether the subject's behaviour changes as a result of exposure to this relationship. In an experiment on classical conditioning, of the type pioneered by Pavlov (1927), the experimenter presents as E1 a neutral stimulus, such as a light or buzzer, and immediately follows this with a motivationally significant E2, such as food or electric shock. The illumination of the light signals the delivery of food, and the experimenter records changes in the subject's behaviour to the light. In Pavlov's terminology, the light is the conditional stimulus or CS, and the food the unconditional stimulus or US. The food unconditionally elicits a set of consummatory responses, one of which is recorded by the experimenter and designated the unconditional response or UR. In Pavlov's case, of course, the UR recorded was the response of salivating. After a number of pairings of CS and US, the CS also starts eliciting salivation, and this is then termed the conditional response or CR. The presentation of the US following the CS is said to reinforce this conditional reflex of salivating to the CS. By extension, therefore, the US is often referred to as a reinforcing event or reinforcer. Pavlov's procedure is known as classical or Pavlovian conditioning.

In an experiment on instrumental or operant conditioning, of the type pioneered by Thorndike (1911) and extensively studied by Skinner (1938), the delivery of the motivationally significant E2 or reinforcer depends on the subject's performance of some designated response as E1. In Thorndike's puzzle box, the animal obtained food only by depressing a particular catch or panel which opened the door and let it out. In the modern equivalent of the puzzle box, the Skinner box, the rat obtains food only by pressing a lever that protrudes from one wall of the box; the depression of the lever (E1) automatically results in the delivery of a small pellet of food into a magazine (E2). In another version of the Skinner box, designed for birds, a pigeon is required to peck an illuminated disc mounted on one wall, and successful pecks result in the automatic delivery of grain to a magazine below the disc. In all these cases, the experimenter records changes in the probability of elicitation of the designated instrumental response over trials or over time.

In both varieties of conditioning experiment, therefore, the experimenter pairs particular events, stimuli or responses with reinforcers, and associative learning may be inferred if it can be shown that the animal's behaviour changes as a consequence of its exposure to the pairing of E1 and E2. In other words, it is necessary to employ control procedures to rule out the possibility that the change in behaviour occurred for some other reason. The phenomenon of sensitisation illustrated one such possibility.

In order to establish, for example, that classical conditioning depends on exposure to a temporal relationship between CS and reinforcer, it is necessary to show that animals exposed to both CS and reinforcer, but without this close temporal relationship between them, do not show similar changes in behaviour. One procedure, used in Evans' experiment with nereid worms, is to present CS and reinforcer alternately, separated by relatively long intervals. As a variant on this, the experimenter may systematically increase the temporal interval separating E1 and E2. In experiments on eyelid conditioning in rabbits, for example, where a CS, such as a tone, is paired with a US consisting of a puff of air directed towards the eye, the CR of blinking to the tone develops reliably only if the interval between the onset of the CS and that of the US is less than two seconds (Schneiderman & Gormezano 1964). If longer intervals between E1 and E2 result in virtually no conditioning, although the animals have certainly been exposed to the two events, one can conclude that it is the close temporal relationship between the two that is responsible for conditioning. Although the temporal interval over which conditioning is possible varies very widely across different procedures (see section 6.3.1), it is always true that increases in the interval beyond a certain limit result in a decline in conditioning, and it is this fact that justifies the assumption that conditioning is a matter of associating events that occur close together in time.

5.3.2 *Laws of associative learning*

Traditional theories of conditioning stressed the importance of temporal contiguity between the events to be associated, to the exclusion of all other factors. They thus reduced conditioning to a very simple process of associating paired events, and rendered it both uninteresting to study and of rather little value to the animals being conditioned. Research over the past 10 to 15 years has established that this picture is grossly oversimplified. Conditioning does not occur automatically just because a CS is followed in time by a reinforcer, for successful conditioning requires that the CS provide information about the occurrence of the reinforcer that is not otherwise available. The process of conditioning thus allows animals to discover what events in their environment predict or cause the occurrence of motivationally significant events like reinforcers; it enables them to distinguish between chance coincidence and true causal relation (Dickinson 1980).

Consider the following experiment. A rabbit receives conditioning trials in which a light, serving as a CS, is paired with the delivery of a US that elicits an eyeblink. The light predicts the occurrence of the US and conditioning occurs. Not surprisingly, reliable conditioning occurs even if the US is presented on only a random 50% of trials, for it still remains true that the onset of the light predicts that the US will probably occur and it is better to be safe than sorry. Finally, one would not expect to see much interference in conditioning to the light if it was accompanied on each trial by one or other of two different tones. The two tones are no better correlated with the delivery of the US than is the light, and provide the rabbit with no better information about what will happen next. All this is confirmed (Wagner *et al.* 1968). As can be seen in the left-hand panel of Fig. 5.3, animals receiving alternating trials to T1–L and to T2–L, on each of which the US occurs 50% of the time, show a high level of conditioning to the light, as revealed by the outcome of test trials to the light alone. (They also

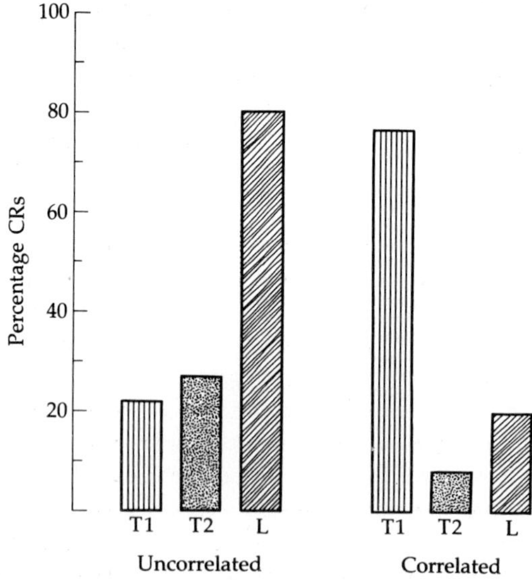

Fig. 5.3. Responses on test trials to individual stimuli, T1, T2 (the two tones) and L (the light), following conditioning to the T1–L and T2–L compounds. In the uncorrelated condition, both T1–L and T2–L were reinforced on a random 50% of trials; in the correlated condition, T1–L was always reinforced and T2–L never. (From Wagner *et al.* 1968.)

show good conditioning to both T1 and T2, but that seems equally reasonable.)

It is quite possible, however, to ensure that the tones do provide significantly better information than the light. If the 50% schedule of reinforced and non-reinforced trials is maintained, but T1 always accompanies the light on trials when the US is presented and T2 always accompanies the light on trials when there is no US, then the tones provide complete information about the outcome of each trial and the light becomes completely redundant. As can be seen in the right-hand panel of Fig. 5.3, the rabbit behaves appropriately, blinking to T1 alone but not to T2, and showing no interest in the light whatsoever. The tones are said to have 'overshadowed' the light. This may seem reasonable and unsurprising and so, no doubt, it is. But the result is important, for it is easy to see that the change in the correlation between the two tones and the US has in no way altered the temporal relationship or degree of correlation between the light and the US. Since the level of conditioning to the light changes markedly even when its relationship to the US is held constant, it follows that conditioning must be affected by other factors. The implication is that one of these other factors is the *relative* predictive value of the CS.

The results shown in Fig. 5.3 form part of a larger pattern. Thus in the experiment of Wagner *et al.* it was the relative correlation of the light and the US that varied. Changes in relative temporal proximity to a US also affect conditioning. In experiments on food aversions (discussed at greater length in section 6.3.1), rats are given a novel, distinctively flavoured substance to eat or drink and are shortly thereafter made mildly ill (for example by an injection of lithium chloride). When they have recovered from their illness, they show evidence of conditioning by refusing to touch the food they had been given before their illness. Revusky (1971) showed that conditioning in rats occurs to the most recently sampled novel food or drink. Animals given saccharin-flavoured water to drink 75 minutes before a lithium injection formed a strong aversion to the saccharin. But if they drank another novel solution (vinegar-flavoured water) in the interval between drinking the saccharin and receiving the injection, they showed virtually no aversion to saccharin. They attributed their illness to the vinegar rather than to the saccharin and, although the temporal relation between saccharin and illness was identical to that experienced by the first group, the presence of a better predictor of their illness, in the form of a novel substance they had ingested more recently, was sufficient to prevent conditioning to the saccharin.

Exactly the same pattern of results has been observed in experiments on instrumental conditioning. For example, pigeons will learn to peck a disc, briefly illuminated with red light, in order to obtain food. They are quite capable of learning to do this even if there is an interval of several seconds between their pecking response and the actual delivery of food. But the occurrence of some other event in this intervening interval completely overshadows this instrumental conditioning: the pigeons attribute the delivery of food to this other event which more immediately precedes it, rather than to their response of pecking the red disc several seconds earlier (Williams 1978). Similarly, rats will learn to run in a running-wheel in order to obtain food, and will do so reliably even if their running response is followed by the delivery of food on only 50% of trials. But if the occurrence of the food is better predicted by a tone which occurs only on reinforced trials (just as the occurrence of the US was better predicted by the tone than by the light in the experiment performed by Wagner *et al.*), then the rats fail to learn the connection between running and food, even though the correlation between the two events remains the same as before (Mackintosh & Dickinson 1979).

In all these experiments, conditioning occurs selectively to a good predictor of the occurrence of a US or reinforcer at the expense of a poorer predictor, and this suggests that the conditioning process is responsible for associating events which are genuinely (causally) related to one another in the real world, rather than events which merely happen occasionally to occur together. Just how this selectivity comes about is not clear. According to one theory, a reinforcer that is already predicted by one event loses its ability to reinforce further conditioning (Rescorla & Wagner 1972). Other theorists have seen a similarity between habituation and selective association (Mackintosh 1975; Wagner 1976; Pearce & Hall 1980); as in experiments on habituation, so in experiments on conditioning, animals learn to ignore a repeatedly presented stimulus if it is not followed by any further, untoward consequence. But regardless of the mechanism involved, the principle of selective association remains of fundamental importance for an understanding of conditioning.

5.3.3 Learning and performance: theories of reinforcement

We have treated conditioning as the prime example of simple associative learning, but this is to ignore what is perhaps the most important fact about conditioning, namely that it results in characteristic changes in behaviour. We can infer that conditioning has occurred only if we observe certain changes in behaviour, but it is also important to understand why particular changes occur and not others. Why does Pavlov's dog salivate to the CS paired with food, or the rabbit blink to the CS that precedes the airpuff? Why, come to that, does the rat in the Skinner box press the lever? Such behaviour depends on learning, but we also need a theory that explains how particular kinds of learning produce particular changes in behaviour. We need a theory of reinforcement.

In fact, we probably need two such theories, one for classical, the other for instrumental conditioning. In instrumental conditioning, Thorndike (1911) argued, responses increase or decrease in probability because of their consequences. This was the celebrated law of effect. A modern version of the law would state that responses associated with appetitive or beneficial consequences increase in probability, while those associated with aversive consequences decrease. This no doubt leaves problems of defining which consequences are appetitive and which aversive, but the more interesting question is whether the law of effect provides a sufficient principle of reinforcement for all cases of conditioning.

Why does Pavlov's dog start salivating to the ticking of the metronome that precedes the delivery of food? It is certainly possible to explain this behaviour in terms of the law of effect, for one could argue that dry food is perhaps unpalatable and its taste and digestibility are improved if it is thoroughly moistened with saliva. The dog learns that salivating in advance of the delivery of food is followed by beneficial consequences, and does so accordingly. Although plausible, this account is almost certainly wrong. Pavlov himself proposed a quite different, and rather simpler, explanation. Salivation, he argued, is a reflex response, naturally or unconditionally elicited by a particular US, dry food in the mouth. As a result of conditioning this US is associated with a CS that regularly precedes it. Pavlov's principle of reinforcement, known as stimulus-substitution theory, says simply that a CS associated with a US may come to substitute for it, and itself elicit those responses normally elicited by the US alone. The dog does not salivate because it has learned that salivating has certain consequences, it is simply that the presentation of a stimulus associated with food in the mouth arouses a representation of the food, and thereby comes to elicit the responses naturally elicited by food.

If salivation were modified by its consequences in the manner suggested by the law of effect, it should be easy to teach an animal to salivate or to stop salivating at will. In practice it is not. A particularly clear example of the insensitivity of Pavlovian CRs to their consequences is provided by the case of eyelid conditioning in rabbits. Why do rabbits blink in anticipation of the aversive US? Is it, as the law of effect implies, because this response reduces the pain or unpleasantness induced by the US? If this were true, rabbits should learn to blink all the more readily if blinking actually caused the cancellation of the US that would otherwise have occurred on this trial. Conversely, they should learn to stop blinking if a cruel experimenter arranged to deliver an extra-severe US on those trials on which they did blink. In fact, they do neither (Coleman 1975). The cancellation of some USs simply leads to an overall decline in the probability of blinking, while the addition of some USs more severe than normal merely serves to increase the probability of responding. Both results are exactly what Pavlov's principle of stimulus substitution would lead one to expect, but completely contradict the law of effect. Punishing the rabbit for blinking merely serves to make it blink all the more.

The rabbit's behaviour seems maladaptive. And so, in a sense, it is, precisely because it violates the law of effect. The adaptive significance of instrumental conditioning is so obvious as to need little comment. If the only way for a hungry rat to obtain food is to perform a particular response such as pressing a lever, then an increase in the probability of lever pressing becomes a prerequisite for survival. The law of effect, indeed, has often been viewed as an analogue of natural selection as a means of shaping the individual's behaviour to the requirements of the environment. The adaptiveness of classical conditioning, however, has sometimes been questioned. Since the dog will get food on each trial regardless of its behaviour, what is the point of salivating just before the delivery of food? Indeed, if classically conditioned responses differ from instrumentally conditioned responses precisely in not being readily modified by their consequences, does this not imply that they are normally without beneficial consequence? It does not. Part of the mistake is to suppose that the particular discrete responses typically recorded by experimenters, such as salivation in response to food or eyeblink in response to a puff of air, exhaust the range of classically conditioned behaviour. As Pavlov fully realised, a dog will do very much more to a CS signalling the delivery of food than just salivate; for example, it will show excitement as soon as the CS is turned on, and approach the place where either the CS is presented or the food delivered. For purposes of scientific analysis, Pavlov chose to ignore these other patterns of behaviour and rely exclusively on conditioned salivation. (Among its virtues, it was easily measured and quantified, and its very arbitrariness discouraged the investigator from making anthropomorphic assumptions about the dog's thoughts and expectations.) But to appreciate the significance of classical conditioning, one must forego this concentration on one arbitrary aspect of conditioned behaviour, and take a wider view of the range of responses that are actually conditioned.

Reinforcing stimuli, such as food or water, elicit a wide variety of responses in hungry or thirsty animals, ranging from general approach behaviour to specific consummatory responses. Similarly, aversive reinforcers, such as electric shocks or attacks by predators, elicit a characteristic array of defensive reactions, which in different species may range from escaping, fleeing and freezing to mobbing and attacking. It does not require great imagination to see that if it is advantageous for a particular animal to curl up into a ball, or to flee when a predator attacks, it will be even more advantageous to do so in response to a stimulus that has regularly preceded attack. Similarly, just as animals must approach food in order to survive, so it must be advantageous to approach stimuli or places which have been associated with finding food.

Classical conditioning, therefore, by ensuring that animals respond to stimuli correlated with the occurrence of reinforcing events in the same way that they respond to those events themselves, often enables animals to behave appropriately in anticipation of a reinforcer, and thus increases the probability of obtaining appetitive reinforcers and evading aversive reinforcers. In this sense, the process of classical conditioning, requiring only the capacity to detect correlations between external events, may be an effective means of producing adaptive modifications of behaviour.

But it does not guarantee success. A general tendency to approach the location of food may not be sufficient actually to obtain food. The predator that merely approaches its prey will find the prey disappearing. Even when obtained, the prey may be inedible without fairly intricate manipulation. Many animals have to learn how actually to obtain their food. A common observation is that a rough approximation to the required behaviour appears in the absence of any opportunity for learning, but that reinforced practice may be necessary to perfect the necessary skill. Young squirrels are able to crack open hazel nuts at the first opportunity, but they gnaw inefficiently and at random until the nut breaks open by chance, and it is only with practice that they learn to gnaw a long furrow down one side, wedge their teeth into the crack, and break the nut in half (Eibl-Eibesfeldt 1963).

What is happening here is that a process of instrumental conditioning is serving to modify the animal's natural appetitive and consummatory behavour so as to maximise efficiency in obtaining the appropriate reinforcer*. The sight of a stimulus associated with food elicits, by a process of classical conditioning, an appropriate set of consummatory responses, but aspects of these responses may then be modified as the animal learns the correlation between variations in its behaviour and variations in its success. Both processes are adaptive, for the opportunity for successful instrumental conditioning would probably not arise without appropriate classical conditioning in the first place, while it is self-evident that instrumental conditioning serves to increase the animal's efficiency. The important point to remember is that although these processes have been studied almost exclusively in the laboratory, they have significant practical value in the real world also.

5.4 Complex learning

Experimental psychologists studying learning in animals have concentrated on conditioning, sometimes to the exclusion of all

*Analogous processes occur when the reinforcer is aversive and the animal's task is to escape or avoid it. As Bolles (1970) has argued, animals have a range of species-specific defensive reactions brought into play by danger or the threat of danger. The process is one of classical conditioning, but the animal may need to select, via instrumental conditioning, which reaction (if any) is most successful in any particular situation.

other possibilities. A chapter such as this must inevitably reflect that emphasis, but it does not follow that all learning can be reduced to conditioning. The ways in which birds learn the characteristics of their species' song (Chapter 3) and the phenomena of filial and sexual imprinting (Chapter 4) suggest that, however important it may be, conditioning does not exhaust the range of possibilities. There have always been psychologists who would have endorsed that conclusion.

5.4.1 Insight

Classical and instrumental conditioning are often said to be instances of simple associative learning; the implication is that there must be other, more complex forms of learning. The Gestalt psychologist Köhler (1925) was one of the earliest proponents of this view, arguing that the behaviourists' analysis of learning was crude and mechanistic, lent plausibility only by the artificial simplicity of the situations in which they studied their animals. For Köhler, learning was largely a matter of perceptual reorganisation. An animal learned to solve a problem when it learned to perceive the constituent elements of that problem in a new light. Isolated at an anthropological field station on Tenerife during the First World War, Köhler studied a group of captive chimpanzees, setting them various problems to solve. The typical problem required them to obtain food lying out of reach, outside the bars of their cage, or suspended from a hook in the ceiling. They learned to use bamboo poles as rakes and, when one was too short, to fit two together to form a longer rake. They climbed up on to boxes to reach the banana hanging from the ceiling and, if one box was too small, would stack one box on another to gain a higher platform. According to Köhler, they were perceiving these objects in a new light, the poles as rakes, the boxes as the elements of a crude ladder.

Although Köhler's chimpanzees were obviously solving problems and doing so quite ingeniously, the evidence of insight and intelligence becomes rather less impressive in the light of further information about the natural behaviour of chimpanzees. Köhler observed his chimpanzees attentively and for long periods of time (far more indeed than most psychologists observe their rats or pigeons in Skinner boxes), but he had not observed their earlier development. Subsequent studies have shown that young laboratory-reared chimpanzees will, given the opportunity, play with sticks and boxes; they will bite the end off one bamboo pole and fit another into the hollow tube so formed; they will pile up boxes and climb up onto the pile (Schiller 1952). All these activities occur in the course of play, when the chimpanzees are neither solving problems nor attempting to obtain food. Their appearance does not depend on the animals having been trained to obtain food by these methods. However, the efficient solution of the sort of problem set by Köhler *does* depend on the opportunity to develop these skills, if only in the course of play. The implication is that Köhler's chimpanzees may have been simply learning what responses were instrumental in obtaining bananas, and that the responses in question formed part of their natural repertoire. This would appear to conform to the definition of instrumental conditioning: a particular response or pattern of responses is selected because it has, in the past, been associated with certain consequences, and those consequences are now valued because they form sub-goals on the route to obtaining the banana.

5.4.2 *Learning sets*

What Köhler's experiments establish, therefore, is that an animal's past experience can have profound effects on its present ability to solve certain kinds of problems. This conclusion appeared to be nicely supported by the work of Harlow (1949) on learning sets. Harlow trained rhesus monkeys on simple two-choice discrimination problems. The animal was confronted with two different objects placed over two small food wells. Displacement of one of the objects revealed a peanut or raisin in the food well. The other well was empty. The monkey's task was to learn which object concealed the reward. Discrimination can be analysed as a case of conditioning—one object or stimulus is associated with the reinforcer and elicits approach, the other is not and does not. But although this analysis may account satisfactorily for the way in which the monkey solves the first discrimination problem it is set, Harlow did not stop there. Once the monkey had solved one problem, it was set another: two new objects were presented, and the monkey had to learn all over again which one was 'correct', and so on. Harlow's monkeys were presented with several hundred discrimination problems, one after the other, and as they progressed through the series, their behaviour changed. Where originally it had taken them a dozen or more trials to learn which object was correct, after one or two hundred problems, as is shown in Fig. 5.4, their behaviour had changed dramatically. Not knowing which alternative was correct on the first trial of a new problem, they chose between them at random and were correct by chance 50% of the time. But on trial 2, they knew the solution and always chose correctly. In Harlow's phrase, they had learned how to learn, or had formed a learning set.

The change in the monkey's behaviour over the course of a few hundred discrimination problems implies that some new process or processes must have been brought into play. If the animals' behaviour on earlier problems can be understood in terms of conditioning, the implication is that other principles must be appealed to in order to account for the way in which they solve later problems so efficiently. The most persuasive account is that provided by Restle (1958). It is apparent that the monkeys must learn something during the course of training on one problem that

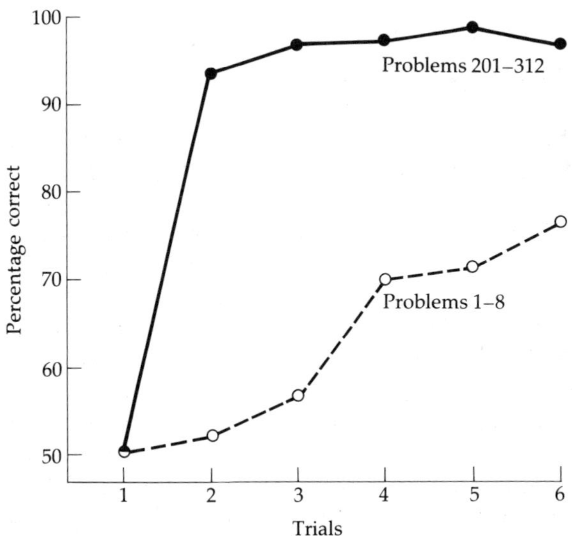

Fig. 5.4. Performance of rhesus monkeys over the first six trials of a series of two-choice discrimination problems. The lower curve represents performance on the first eight problems, the upper represents performance after training on 200 problems. (From Harlow 1949.)

is relevant to the solution of the next. Now although the actual stimuli or objects change from one problem to the next, it is not too difficult to see that there are features of the situation that do remain constant across problems and therefore provide the basis for transfer. For instance, it is always true that the object chosen and rewarded on one trial will be rewarded on the next. If a monkey can learn therefore to characterise the two objects, not simply in terms of their physical features, but also in terms of which one it chose on the preceding trial and what was the outcome of that choice, then it is provided with a basis for choice on the next trial that will remain valid across all problems. In Restle's terms, the monkey must learn a win–stay, lose–shift strategy, learning to repeat its choice on the last trial if that choice was rewarded and shift if it was not.

Restle's analysis does not question the associative basis of performance of learning sets. The experienced monkey is still assumed to be associating one alternative with food and the other with its absence, and choosing between them accordingly. What has changed is the monkey's characterisation of the alternatives. There is good reason to believe that this is often the difference between simple conditioning and more complex forms of learning. The complexity resides largely in the sorts of stimuli to which the animal is responding and the way in which they are analysed or categorised, rather than in any complexity of their relationship to further events of consequence to the animal. The point can be illustrated by two final examples.

5.4.3 *Learning of spatial relationships: rats in mazes*

Ever since Willard Small (1901) first trained rats to run through a miniature replica of the Hampton Court maze, psychologists have studied rats in mazes. Indeed, the image of the psychologist watching his hungry rat trundling its way through a maze to find food in the goal-box has often been held up to ridicule as the perfect example of the futile and absurd experimentalist. The derision is surely misplaced, for rats are remarkably well adapted to finding their way through mazes, and the processes whereby they learn the spatial layout of a maze are far from understood. Interest in the topic of maze learning has shown a marked revival in recent years, partly due to the suggestion that such learning depends on the establishment of a cognitive map of the rat's environment, and that this map is located in the hippocampus (O'Keefe & Nadel 1978).

A simple T-maze is illustrated in Fig. 5.5. The rat is placed in the start-box at the outset of each trial; the correct goal-box is baited with food, and the rat's task is to choose correctly. The problem can be described as a spatial discrimination. The question is: what is the basis on which the rat discriminates between the two goal-boxes? There are two relatively simple solutions, but even when both of these have been ruled out the rat is still capable of solving the problem. One possibility is that the two goal-boxes, and arms leading to them, differ in some intrinsic characteristics; the rat associates one set of characteristics with food and learns to approach them. Although this is a common basis for solution, it can be ruled out by constructing the goal-arms and -boxes of identical materials and, more important, by randomly interchanging them from trial to trial. The rat can still learn to go to the

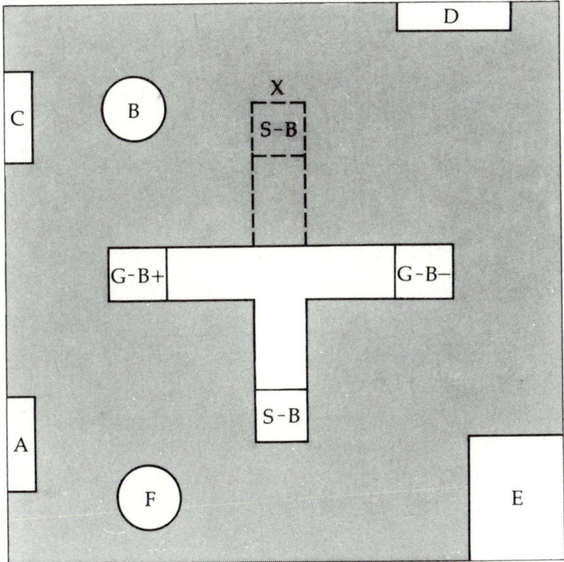

Fig. 5.5. Schematic diagram of a T-maze. S-B is the start-box, G-B+ the goal-box containing food, and G-B– the empty goal-box. A, B, C, etc., represent 'landmarks' located round the maze (e.g. tables, windows, rat cages, etc.). The dotted start-arm and start-box located at X are for a test trial to determine whether the rat has learned to turn left at the choice point or to approach G-B+.

left-hand goal-box to obtain food. The second possibility is that the rat learns to make a left-hand turn at the choice-point. Although rats do sometimes solve a maze in this way, it is a relatively uncommon solution, for if they are given a test trial with the start-arm located on the opposite side of the maze (with the start-box at X), they will still approach the rewarded goal-box, thus making a right-hand turn rather than the left-hand turn which had been rewarded during training (Restle 1957; see also Morris 1981).

Such a solution depends on the existence of a number of landmarks outside the maze, such as those at A, B, C, etc., in Fig. 5.5; in the absence of such external landmarks or internal features differentiating the two goal-boxes, the rat will indeed solve the problem by learning to make a particular turn at the choice-point. Now it might seem that if appropriate landmarks are located immediately behind the two goal-boxes, there is little distinction between this solution and the first one, whereby the rat learns to approach a particular set of stimuli characterising the rewarded goal-box. But in fact it is quite unnecessary for the landmarks to be anywhere near the two goal-boxes; indeed, it is apparent that the rat does not define the two goal-boxes in terms of their each being associated with a particular nearby landmark, but rather in terms of their spatial relationship to a large number of different landmarks, no one of which is indispensible (e.g. O'Keefe & Conway 1978; Suzuki *et al.* 1980). Thus in the present example, the positive goal-box might be characterised as lying at such and such an angle between A on the left and B on the right, and perhaps at a particular distance and angle from E and F. The removal of one of these landmarks would not be sufficient to disrupt performance, but if they were rearranged in such a way as to destroy their former spatial relationship to one another, the rat would be at a loss. Maze problems, therefore, although in a sense examples of very simple conditioning in which one of two alternatives is associated with food and therefore approached, clearly bring into play a complicated sense of spatial relationships whereby the rat defines those alternatives.

5.4.4 *The learning of second-order relationships*

It is clear, then, that animals can respond to quite complex relationships between events, and it is worth concluding with examples of the consequences of such an ability. In a simple discrimination problem, an animal is confronted with two alternatives, for example a red and a green light, one of which is consistently associated with reinforcement, the other not. Animals ranging from chimpanzees to pigeons can also be taught conditional discriminations, where choice of red is correct on some trials, and choice of green is correct on others. Which alternative is associated with reinforcement on a given trial depends on the value of a conditional cue: the presence of one tone, for example, signals that red is correct, while the presence of another signals that choice of green will be reinforced.

A special case of such a conditional problem is when the conditional cues are the same as the alternatives between which the animal must choose. In a so-called 'matching-to-sample' task, a red sample stimulus signals that the red alternative is correct, while a green sample signals that green is correct. In an 'oddity' problem, a red sample would signal that green was correct, and a green sample would signal that red was correct. One may now ask how animals solve such problems. Do they solve them as they would any other conditional discrimination, learning that a particular conditional cue signals a particular stimulus–reinforcer relationship? Or do they learn the more general principle, that the correct alternative is the stimulus that is the same as (or different from) the sample? Are they, in other words, capable of responding in terms of the abstract relationship of similarity or identity? The decisive test of such a capacity is to see whether the animal shows transfer to a new set of stimuli—whether animals that have learned to select the alternative that is the same as the sample will continue to do so regardless of the actual physical characteristics of the stimuli. Both rhesus monkeys (e.g. Mishkin & Delacour 1975) and dolphins (Herman & Gordon 1974) are capable of solving such problems when the actual stimuli are changed on every single trial. In their case, therefore, there is no question but that they are responding in terms of the relationship between the sample and alternatives. Pigeons show some evidence of transfer (Zentall & Hogan 1974), but is is so slight that considerable doubt, probably well justified, has been raised as to whether they are actually responding to the relationship (Carter & Eckerman 1975). Nevertheless, the fact that at least some animals are capable of responding in terms of this sort of relationship between stimuli implies perceptual or cognitive capacities not readily captured by simple associative analyses. As Premack (1978) has argued, such relational responding is probably a necessary precursor for learning even the rudiments of any linguistic problem.

5.5 Conclusions

Psychologists and zoologists have studied learning in a variety of animals and in an even wider variety of situations. The present survey has only touched on some of this diversity, but it may have been sufficient to give the reader pause before attempting to propose a single monolithic theory of learning. The days of such theories, even in psychology, are long past. It does not seem probable that habituation and language learning have very much in common.

However, the differences should not be exaggerated. The important question is whether the same processes or mechanisms are operating even in quite diverse situations. It is often possible that they are. If habituation involves animals learning that a particular, regularly repeated stimulus has entirely predictable or entirely unimportant consequences, then such a learning process will quite certainly also be engaged in experiments on conditioning. After conditioning has been achieved, a CS is a stimulus with predictable consequences, while the irrelevant stimulus in a study of overshadowing may be defined as one with unimportant, because otherwise predicted, consequences. Similarly, although there is no doubt a large gap between what is happening when Pavlov's dog starts salivating at the ticking of the metronome and the behaviour of a rhesus monkey solving one-trial matching-to-sample problems, the performances of both animals probably depend on related associative processes.

Nevertheless, it is easier to be impressed by the diversity of things animals can learn than by their similarities. All the examples considered here have conformed to the definition of learning outlined in the introduction: one can infer that an animal's change in behaviour is due to learning when one can show that a particular set of circumstances was responsible for that change. But neither the changes in behaviour nor the circumstances are easily reduced to a single formula.

5.6 Selected reading

The theory of conditioning, largely from an associationist point of view, is discussed by Dickinson (1980), Rescorla (1975) and Mackintosh (1974). Two other books on animal learning, written by psychologists but from a point of view more influenced by B.F. Skinner, are those by Schwartz (1978) and Davey (1981). Kandel (1976) gives an account of the neurophysiology of learning, with particular emphasis on habituation. An account of learning written from a more ethological perspective is provided by Manning (1979).

CHAPTER 6 LEARNING AS A BIOLOGICAL PHENOMENON

T.J. ROPER

6.1 Introduction: psychological and ethological approaches to learning

In most areas of ethology there has been a close association between a naturalistic approach in which behaviour is observed and an analytical approach in which experiments are conducted. At the same time, ethology has usually dealt both with questions about the causal mechanisms of behaviour and with questions about the biological functions of behaviour. In the case of learning, however, ethological interest has remained mostly at a naturalistic level, and has dealt primarily with questions about function. Consequently, a wealth of information has been amassed about the ability of particular species to learn particular things but, with a few important exceptions (imprinting, song learning, habituation), these instances of natural learning have not been subjected to detailed analytical scrutiny.

Meanwhile psychologists, working from a radically different intellectual standpoint, have developed their own approach to learning. 'Learning theory', as it is known, is highly analytical, being characterised by an exclusive concern with the mechanisms of classical and instrumental conditioning. Its methodology involves laboratory tests of learning in a few species (rat, pigeon and monkey), in contrived environments (maze, Skinner box, classical conditioning apparatus).

Learning theory has remained a closed book to most ethologists whilst, conversely, learning theorists have usually ignored the ethological literature; each side has felt that the approach of the other is irrelevant to its own theoretical concerns. In the late 1960s, however, some psychologists began to suspect that learning theory had given rise to an oversimplified view of learning, and that a more naturalistic approach should be adopted, an opinion always held by ethologists. The reason for this crisis of confidence amongst learning theorists was the emergence of various lines of evidence, known as 'constraints-on-learning' evidence, which apparently contradicted two fundamental principles of learning theory. The threatened principles became known, respectively, as 'general process theory' and 'the principle of equipotentiality'.

The 'contraints-on-learning' evidence seemed to presage a fundamental revolution in psychological thought which would at last reconcile the psychological and ethological approaches to learning. More recently, however, traditional learning theory, and in particular the view known as 'general process theory', has begun to make a comeback, and now it seems that learning theory is in some respects as far removed from ethological interests as it ever was.

The main purpose of this chapter is to review the constraints-on-learning evidence and to examine its impact on our view of learning as an adaptive phenomenon. Sections 6.2 to 6.4 deal with the evidence for constraints on learning and its implications for psychological learning theory, while subsequent sections deal with the ethological approach to learning. Before turning to the actual constraints-on-learning evidence I shall first give a brief account of traditional learning theory, and in particular of general process theory and the principle of equipotentiality.

6.2 The psychological approach: traditional learning theory

Here I shall try to convey some idea of the traditional learning theory approach, so that the reader will know what it is that the constraints-on-learning evidence was supposed to challenge. This section also introduces some essential terminology.

6.2.1 Procedures of classical and instrumental conditioning

Classical and instrumental conditioning are often known as associative learning, because they are thought to involve a learned association between two events. Let us call these events E1 and E2 (Dickinson 1980). In a conditioning experiment the animal experiences a series of 'trials', in each of which E1 and E2 occur with a particular temporal relationship to one another; typically, E2 follows immediately after E1 (a procedure known as *pairing* of the two events). Learning is inferred if, as a consequence of this experience, some aspect of the animal's behaviour changes (see Chapter 5, especially section 5.3.1).

In classical conditioning, E1 is a stimulus presented by the experimenter (e.g. a bell or a flashing light), and E2 is a motivationally significant event or *reinforcer* (e.g. food, water or an electric shock). The measure of learning is a change in the occurrence of the behaviour originally elicited by E2. For example, in Pavlov's famous experiments, a bell (E1) was repeatedly followed by presentation of food (E2) to a dog. As a consequence of being paired with food, the bell came to elicit a response (salivation) which was previously elicited only by food.

In instrumental conditioning E1 is not a stimulus but a *response* made by the animal, for example pressing a lever or turning left in a maze. (This difference in the nature of E1 is the feature which distinguishes instrumental from classical conditioning for the purposes of definition.) In instrumental conditioning, as in classical conditioning, E2 is a reinforcer, but the measure of behaviour in instrumental conditioning is the frequency or intensity of the response E1 itself. For example, in Skinner's experiments a rat was given food (E2) every time it pressed a lever (E1); as a consequence of this pairing of lever-pressing with food, the rate of occurrence of lever-pressing increased. Because the animal has to find out for itself that a particular response results in delivery of a reinforcer, instrumental learning is colloquially known as 'trial-and-error' learning.

6.2.2 *General process theory*

General process theory is the view that *all instances of associative learning involve the same basic underlying mechanism or process.* This idea, although rarely stated explicitly in textbooks, is crucial to learning theory in two respects. First, it provides learning theory with its ultimate goal, namely to discover the nature of the basic learning process. Secondly, general process theory provides a rationale for the laboratory-based methods of learning theory, via the argument that if a common underlying process of learning exists then it is only likely to be revealed by artificial test procedures in which species-specific behaviour patterns are minimised (e.g. Dickinson 1980). In other words, learning theorists confine their work to the laboratory not because they are too lazy to observe the behaviour of animals in the wild, but because they believe that *only* in the laboratory will the hypothetical general learning process manifest itself in a recognisable form. By the same token, learning theorists can justify their predilection for rat and pigeon because, if the learning process is believed to be essentially the same in all species, then the choice of experimental animal can be based solely on considerations of practical convenience.

What do learning theorists mean when they say that all instances of associative learning involve the same basic 'mechanism' or 'process'? A naive reader might assume that the term 'mechanism' refers to the neural and physiological hardware responsible for the changes in behaviour that we call learning. But this interpretation would be quite wrong; learning theorists have traditionally refused to investigate mechanisms at a neural level, claiming that behavioural explanation should precede physiological analysis. Instead, they have sought to specify the formal relationships between input variables (concerning the nature of events E1 and E2 and the manner in which they are paired) and output variables (measures of behaviour) in the *intact* animal. Consequently, when a learning theorist speaks of a particular mechanism or process of learning, what he means is a relationship between experience and behaviour that can be described by a particular set of formal rules or 'laws' of learning.

General process theory, then, translates into the view that *all instances of associative learning obey the same basic laws.* This, of course, leads us to the question: what constitutes a 'law' of learning?

6.2.3 *Some examples of laws of learning*

No psychology textbook contains a list of 'laws of learning' as such—learning theory is too contentious a discipline. Nevertheless, it is possible to formulate various statements which, until the advent of the constraints-on-learning evidence, were assumed to apply to all instances of associative learning. Below are a few concrete examples of such statements, selected primarily for illustrative purposes. For detailed references to the experiments on which these and other 'laws' are based, see Kimble (1961) and Mackintosh (1974).

(1) *Associative learning only occurs when the time interval between E1 and E2 is short.* Many early conditioning experiments suggested that for associative learning to take place E1 must precede E2 by a time interval lasting not more than a few seconds, or at the most a few minutes (Fig. 6.1).

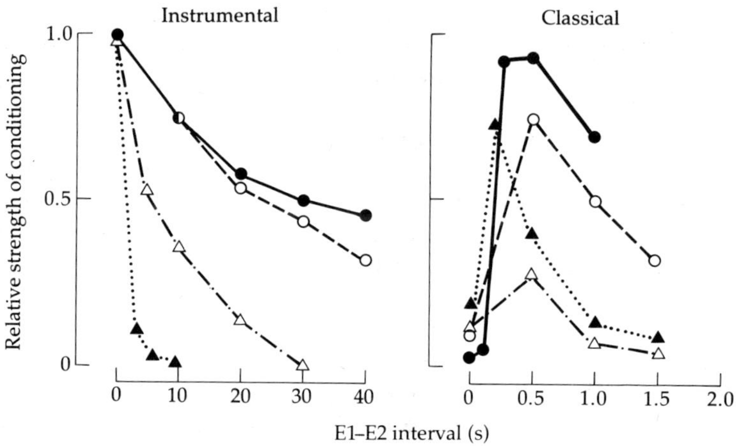

Fig. 6.1. Strength of learning as a function of E1–E2 interval in four studies of instrumental conditioning (left) and four studies of classical conditioning (right). Instrumental conditioning is maximal with zero E1–E2 interval, and rarely survives intervals of more than about 60 s; classical conditioning is maximal with intervals of about 0.5 s, and rarely survives intervals of more than about 2.0 s. (After Kimble 1961.)

(2) *In classical conditioning the speed or strength of learning increases with the intensity of the stimulus E1 (within limits).* Again, many early experiments support this generalisation. For example, Pavlov found that a dog would learn to salivate more rapidly to a loud buzzer than to a soft buzzer when these stimuli were paired with food.

(3) *In associative learning the speed or strength of learning increases with the size of the reinforcing event E2 (within limits).* For example, a rat will learn to run a maze faster, the larger the food reward.

(4) *When the reinforcer is witheld, the learned response declines in frequency and/or intensity.* For example if, after conditioning has taken place, food is no longer presented every time event E1

occurs, a classically conditioned response such as salivation or an instrumentally conditioned response such as lever-pressing will gradually cease to occur. This gradual cessation of responding is known as *extinction*.

(5) *When E1 is a compound stimulus whose elements differ in intensity, conditioning occurs only to the more intense element*. If, in a classical conditioning experiment, E1 consists not of a unitary stimulus (e.g. a buzzer) but of a compound stimulus (e.g. a buzzer plus a light), Pavlov and others have shown that only the more intense stimulus element in the compound stimulus comes to elicit the learned response. The more intense element is said to *overshadow* the less intense element.

Overshadowing is an important phenomenon because it shows that mere pairing of a stimulus with a reinforcer does not guarantee learning. To put it another way, overshadowing shows that conditionability is not an absolute property of a given stimulus, but depends on the presence or absence of competing stimuli (see section 5.3.2).

(6) *Prior conditioning of one element of a compound stimulus prevents subsequent conditioning of other elements*. In a normal classical conditioning experiment, a stimulus A (e.g. a red light) is paired with a reinforcer (e.g. food). Consequently, A comes to elicit a response (salivation). Suppose that, subsequently, stimulus A plus another stimulus of equal intensity, B (e.g. a green light), are presented together, and are again followed by the same reinforcer (food). The result is that stimulus B does *not* come to elicit the response of salivation, despite the fact that B has been paired with food. This is known as *blocking*, prior conditioning to A being said to block conditioning to B.

Like overshadowing, blocking shows that mere pairing of stimulus and reinforcer is not a sufficient condition for learning. In the case of blocking, learning depends on whether or not the reinforcer in question has previously been paired with another stimulus; that is, it depends on the animal's previous history of learning.

6.2.4 *The principle of equipotentiality*

In the most extreme form, the principle of equipotentiality states that *all pairs of events E1 and E2 can be associated with equal ease, in any species*. Thus, the principle of equipotentiality implies that any animal can learn virtually anything. According to this view, when learning fails to occur, as obviously it often does (for example we cannot teach a rat to recite Shakespeare), this is because of limitations on the species' sensory or motor capacity, rather than because of limitations on its ability to learn.

To ethologists the principle of equipotentiality has always seemed absurd, because there are many examples of natural learning being limited by species-specific factors (see section 6.5). Prior to the constraints-on-learning debate, however, most psychologists ignored the ethological evidence or explained it away (see Seligman 1972; Garcia *et al.* 1972; Shettleworth 1972 for documentary evidence of the prevailing attitude of psychologists).

Why did the principle of equipotentiality become so firmly entrenched in the dogma of learning theory? Initially there was some empirical evidence in its favour; Pavlov and Thorndike were surprised to find that unnatural stimuli such as lights and buzzers, and unnatural responses such as operating a mechanical latch, could serve as a basis for classical and instrumental conditioning. Subsequently, however, the idea that animals and humans could learn anything became inextricably associated with the view that the mind is a blank slate, written upon indiscriminately by experience but immune from the influence of instincts (the so-called *tabula rasa* doctrine of behaviorism, most notoriously espoused by Watson (1930)). The idea of equipotentiality was therefore a vital element in the behaviorists' doctrinaire stand against instinctivism; it became an article of faith rather than a scientific hypothesis. (For an account of the historical background to the link between behaviorism and environmentalism see Boakes, in press.)

6.2.5 *General process theory and equipotentiality distinguished*

Most people who have written about constraints on learning have implied that general process theory and the principle of equipotentiality amount to much the same thing. This is wrong. General process theory does not imply that animals can learn anything, it merely implies that *when associative learning occurs* it will obey certain laws (such as those stated in section 6.2.3). The principle of equipotentiality *does* imply that animals can learn anything, but it does not necessarily imply that the laws of learning will be the same in all cases, though in practice this was usually assumed to be so. The distinction is important, because it means, as we shall see, that evidence contrary to the principle of equipotentiality does not necessarily undermine general process theory. With that in mind, we can now turn to the actual evidence concerning constraints on learning.

6.3 The constraints-on-learning evidence

The constraints-on-learning evidence was thought to show (a) that not all examples of associative learning obey the same laws, and (b) that animals do not learn all tasks with equal facility. The phrase 'constraints on learning' was intended to imply that learning, rather than occurring indiscriminately, is subject to species-specific and task-specific limitations and predispositions.

The constraints-on-learning evidence was mostly experimental in nature and arose within the learning theory tradition. Hence, unlike ethological evidence with similar implications (see section 6.5), it was taken seriously by psychologists. (For more detailed reviews see Bolles 1970; Seligman 1970; Seligman & Hager 1972; Shettleworth 1972; Garcia *et al.* 1973; and Hinde & Stevenson-Hinde 1973.)

6.3.1 Food-aversion learning

Rats, like humans, develop an aversion to foods that make them ill. This food-aversion learning (which, incidentally, makes rats very difficult to poison) seems to be an example of classical conditioning, with flavour of the food constituting event E1, sensation of illness constituting event E2, and avoidance of the food constituting the measure of learning.

From the point of view of learning theory, however, this analysis poses two problems. First, the rat presumably does not feel ill until some minutes or even hours after eating, so that flavour (E1) and illness (E2) are separated by a considerable interval of time. How does learning manage to occur despite this temporal gap, given that classical conditioning normally requires a very short E1–E2 interval (cf. Fig. 6.1)? Secondly, why does the rat associate the *flavour* of the food with illness, rather than some other feature of the food such as its appearance or location?

Flavour, appearance and location of the food are all paired with illness to the same extent so, according to the principle of equipotentiality, all should be equally associated with illness. Below we consider in detail first the problem of flavour–illness delay, and then the problem of flavour–illness specificity.

The problem of flavour–illness delay

If a rat is fed poisoned food, it is impossible to know precisely when it first begins to feel ill, and hence to know how long the E1–E2 interval really is. Garcia *et al.* (1966) overcame this problem by inducing illness artificially, thus allowing the time of onset of illness to be manipulated precisely. A rat was allowed to consume a novel food (say, a saccharine solution), after which it was made ill by exposure to X-irradiation or by injection with lithium chloride or apomorphine (treatments which produce the symptoms of food poisoning). When the rat had recovered it was offered the original food again. If the rat rejected the food the second time around, it must have formed a learned association between food and illness.

Several experiments of this type have been carried out, varying the time interval between the rat's first consumption of a novel food and the experimenter's subsequent application of an illness-inducing treatment. The results show that food-aversion learning

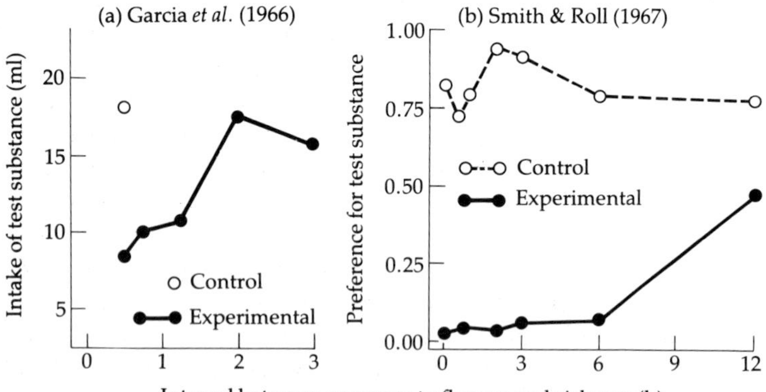

Fig. 6.2. Strength of food-aversion learning as a function of interval between E1 (taste) and E2 (illness), in two separate experiments. Control subjects were allowed to taste the test food (saccharine solution) but were not made ill. Note that learning involves suppression of consumption of the test food, so that the lower the points on the graph, the stronger is the learning. (After Mackintosh 1974.)

can survive delays of up to 12 h between consumption and illness (Fig. 6.2).

The claim of Garcia and colleagues to have demonstrated conditioning with such unusually long E1–E2 delays caused a furore in the world of learning theory, in which a common reaction was frank disbelief of the results (Revusky 1977b). When it became clear that the results were repeatable, various attempts were made to explain them away, either by arguing that food-aversion learning was not true conditioning (e.g. Bitterman 1975, 1976), or by arguing that there was not really a long E1–E2 interval. For example, it was suggested that an aftertaste of the novel food must persist until the rat became ill. These and similar arguments have subsequently been refuted by experiments using stringent control procedures (for a good review of the methodological issues see Domjan 1980); and it is now generally accepted that food-aversion learning really is what it appears to be, an example of conditioning in which the E1–E2 interval can be up to several hours long. Food-aversion learning therefore violates one of the most firmly established of the empirical 'laws' of conditioning (cf. section 6.2.3).

The problem of flavour–illness specificity

Do rats really have a predisposition to associate the flavour of a food, rather than say its appearance or location, with illness? Garcia and Koelling (1966) allowed rats to drink water which was novel either because of its flavour (saccharine or salt) or because it was wired up so that whenever the rat's tongue touched the water a light flashed and a clicker sounded. Garcia and Koelling described these situations as 'tasty water' and 'bright-noisy water', respectively. Some of the rats which had drunk tasty water were then exposed to X-rays to make them ill, while others were subjected to painful electric shock; similarly, some of the rats which had drunk bright-noisy water were made ill, while others were shocked. After recovery, the rats were again presented with the type of water (tasty or bright-noisy) that they had originally drunk. The result was that tasty water was avoided if it had previously been paired with illness, but not if it had been paired with shock, whereas bright-noisy water was avoided if it had been paired with shock, but not if it had been paired with illness (Fig.

6.3). This experiment suggests, then, that rats readily associate flavour with illness, or exteroceptive cues (light–clicker stimulus) with shock, but they do not readily associate flavour with shock, or exteroceptive cues with illness. Subsequent experiments along similar lines, but incorporating more stringent control procedures, have substantiated this result (see review by Domjan (1980)).

Birds tend to identify noxious prey by its appearance rather than by its taste, so in birds we would expect a predisposition to associate appearance of food, rather than flavour, with illness. Early experiments (e.g. Wilcoxon *et al.* 1971) seemed to confirm this expectation, but it now appears that the situation is not so simple. For example, chickens and quail do tend to associate visual cues with illness, but pigeons and hawks do not; chicks rely on visual cues when food is paired with illness, but on taste cues when water is paired with illness (Gillette *et al.* 1980; Lett 1980).

The precise nature of these puzzling species differences remains to be determined (see also section 6.4.2). At the time of

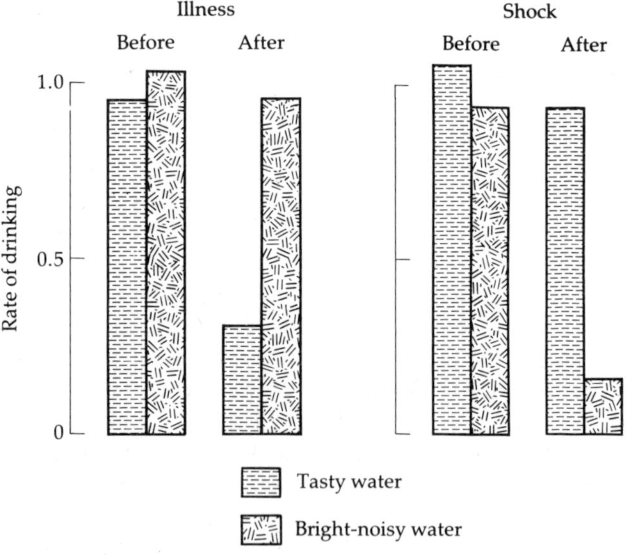

Fig. 6.3. Rate of drinking of tasty water and bright-noisy water before and after exposure to illness-inducing X-irradiation (left) or electric shock (right). Exposure to irradiation caused a reduction in consumption of tasty water only; exposure to shock caused a reduction in consumption of bright-noisy water only. (After Garcia & Koelling 1966.)

writing, however, food-aversion experiments seem to undermine the principle of equipotentiality in two ways. First, they show that in a given species certain E1–E2 pairings result in rapid learning whereas others do not. Secondly, they show that species differ in the types of events that they find easy to associate.

6.3.2 *Cue–reinforcer specificity in instrumental conditioning*

A common procedure in instrumental conditioning is to set up what is known as a discrimination task, in which the learned response is paired with reinforcement only in the presence of a particular stimulus, the *discriminative stimulus*. For example, the response of lever-pressing might be reinforced with food when a red light is on, but not when the red light is off. As a result, the animal learns to restrict its responding to times when the discriminative stimulus is present.

According to the principle of equipotentiality, all detectable stimuli should be equally capable of acting as discriminative stimuli. However LoLordo and others have shown that in pigeons a discrimination with avoidance of shock as reinforcer is learned more easily on the basis of auditory discriminative stimuli, whereas a discrimination with food as reinforcer is learned more easily on the basis of visual stimuli (see review by LoLordo (1979)).

This effect was first demonstrated by Foree and LoLordo (1973) using an overshadowing procedure in which a compound cue of red light plus tone was used as a discriminative stimulus for the response of treadle-pressing. In one group of pigeons treadle-pressing during the light–tone stimulus resulted in food reinforcement; in another group it resulted in avoidance of electric shock; and in neither group did treadle-pressing in the absence of the stimulus produce reinforcement. Consequently, both groups learned to treadle-press only during the light–tone stimulus. Foree and LoLordo then presented either the tone alone or the light alone, to see which of these two elements of the compound stimulus would elicit the response. They found that when food had been the reinforcer, treadle-pressing occurred during the light but not the tone, whereas when shock avoidance had been the reinforcer, treadle-pressing occurred during the tone but not the light. In other words, either the light overshadowed the tone or vice versa, depending on the nature of the reinforcer.

In its design and implications this experiment resembles those described in the previous section, but it shows differential learning of particular cue–reinforcer associations in instrumental rather than in classical conditioning.

6.3.3 *Response–reinforcer specificity: autoshaping*

Perhaps the most important implication of the principle of equipotentiality, at least from the point of view of behaviorism as a socio-political doctrine, is the view that any response in an animal's behavioural repertoire can be strengthened by instrumental conditioning. According to this view the responses actually chosen for laboratory experiments (pressing a lever in the case of rats, or pecking a lighted disc in the case of pigeons) are entirely 'arbitrary', in the sense that any other response could have been chosen instead.

Recent evidence suggests, however, that when a pigeon pecks at a disc for food reinforcement it is not producing an arbitrary response, it is producing species-specific feeding behaviour. The evidence, which results from detailed analysis of videotapes of pecking, shows that the instrumental peck closely resembles a peck at actual food—the peck is forceful but of short duration, the beak is opened at the moment of contact and the eyes are closed. By contrast, when a pigeon pecks a disc for water reinforcement the peck is of relatively long duration, less force is exerted, the beak

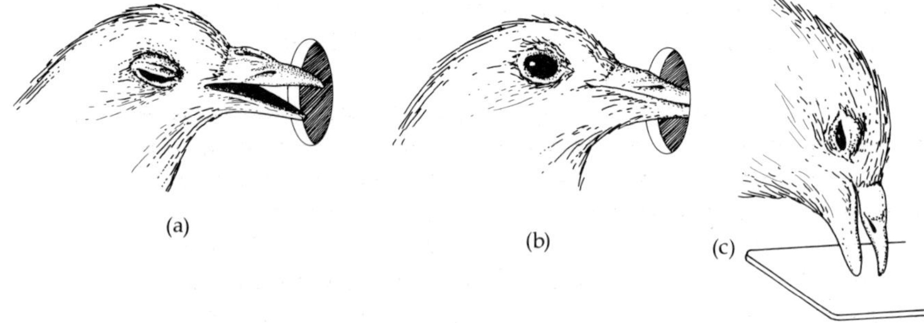

Fig. 6.4. (a) Key-pecking in a pigeon with food as reinforcer. Note the closed eye and open beak, typical of pigeon feeding behaviour. (b) Key-pecking with water as reinforcer. Note the open eye and closed beak, typical of pigeon drinking. (c) An attempt was made to train this pigeon to press a lever with its foot. Instead it pecks the lever. (Based on photographs in Moore 1973.)

is opened only slightly at contact, the tongue is extruded, and the eyes remain open, just as if the bird were ingesting actual water (Fig. 6.4).

This suggests that instrumental learning, in the sense of learning a new response because it leads to reinforcement, is not occurring in the pigeon at all. Rather, the bird acts as if it learns an association between the *disc* and the reinforcer; to put it informally, the bird seems to 'mistake' the disc for actual food or water. If this analysis is correct, then conditioning is classical rather than instrumental, because the learned association is between a stimulus and a reinforcer rather than between a response and a reinforcer.

The idea that 'instrumental' pecking is really a classically conditioned response is strengthened by demonstrations that pecking can indeed be classically conditioned, using what is known as an 'autoshaping' procedure (Brown & Jenkins 1968). A pigeon is confined in a Skinner box, and periodically a small disc is illuminated, followed immediately by presentation of food. Soon the pigeon begins to peck the key, despite the absence of any instrumental contingency between pecking and food presentation. In fact, pecking can even be autoshaped despite a negative instrumental contingency, such that pecking actually prevents food from appearing (Williams & Williams 1969). This latter effect is important, because it more or less rules out the possibility that autoshaped pecking owes its occurrence to covert instrumental conditioning of the type thought to underlie 'superstitious behaviour', rather than to classical conditioning (see Staddon & Simmelhag 1971 and Hearst 1978 for extended discussion of this issue).

The discovery of autoshaping caused immense interest because it questioned the very existence of instrumental conditioning, and thus undermined the very foundation of behaviorism (Schwartz 1981). People began to ask whether the pigeon was capable of instrumental learning at all. Attempts were made to train pigeons to press levers with their feet for food reinforcement, but the pigeons tended to peck the levers instead (Fig. 6.4c). Similarly, when a pigeon was required to peck a key to obtain access to a mate it started to 'court' the key instead of pecking it (Moore 1973). Subsequently, it *has* proved possible to train pigeons to operate levers with their feet, but only with special procedures designed to eliminate the unwanted response of pecking.

Further examples of cases in which unwanted classically conditioned responses interfere with instrumental learning are given by Breland and Breland (1961) and by Sevenster (1973), in a range of species from raccoons to sticklebacks. (The paper by Breland and Breland is especially recommended.) Nevertheless, 'true' instrumental conditioning does seem to occur more readily in rats and other mammals than in pigeons. The archetypal response of lever pressing does not obviously resemble the rat's eating behaviour, though it is true that rats do bite and lick the lever as well as pressing it with their paws. Also, rats respond differently from pigeons when subjected to an autoshaping procedure. Instead of approaching and attempting to 'eat' a stimulus that is paired with food, they tend, as a rational creature would, to remain at the place where the real reinforcer is due to appear (Boakes 1979).

With reinforcers other than food there is often less reason to suspect the influence of classical conditioning, because the response bears less similarity to the consummatory behaviour elicited by the reinforcer. For example, a monkey will press a lever in order to see another monkey (Butler 1953); a rat will drink water in order to run in a wheel (Premack 1959); a human will manipulate a switch in order to turn on a light.

To conclude, it is probably overstating the case to say that there is no such thing as instrumental conditioning of an arbitrary response (though see the article by Moore (1973) for an approach to this radical view). Nevertheless, the existence of autoshaping and related phenomena undermines the principle of equipotentiality by showing (a) that responses are more easily learned if they are compatible with the reinforcer in the sense of containing a classically conditioned component, and (b) that the capacity for acquiring arbitrary instrumental responses differs between species, being greater in mammals than in fish or birds.

6.3.4 *Unconditionable instrumental responses*

So far, we have been discussing constraints on learning that reflect different degrees of compatibility between a stimulus or response on the one hand and a reinforcer on the other. In this section we consider whether there exist responses which are unconditionable in an absolute sense, and in the following section whether some reinforcers are intrinsically weaker than others.

Thorndike (1911), who is credited with the discovery of instrumental conditioning, attempted to train cats to perform various responses in order to escape from confinement in a 'puzzle box'. Some responses, such as pulling on strings or nuzzling latches, were relatively easily learned; but self-directed responses, such as licking or scratching particular parts of the body, were not; during training they tended to become fragmentary and to disappear. Konorski (1967) encountered similar problems when he tried to condition yawning as an instrumental response in dogs.

In a more recent systematic series of experiments Shettleworth (1975, 1978, 1981; Shettleworth & Juergensen 1980) has attempted to condition a wide range of responses (rearing, face-washing, scratching, digging, scent-marking, etc.) in hamsters, with a variety of reinforcers (food, nest material and brain stimulation). Some activities, such as scrabbling, rearing and digging, proved easy to condition with all reinforcers; others, such as face-washing and scent-marking, showed little or no evidence of conditioning with any reinforcer (Fig. 6.5).

There are two possible reasons why a response such as face-washing might fail to increase when it is paired with a reinforcer. One is that there might be a learning deficit—the animal is unable to form an association between response (E1) and reinforcer (E2). The other is that there might be a *performance* deficit—the animal forms the necessary association, but is unable for some other reason to perform the response with greater than baseline frequency.

Morgan and Nicholas (1979) devised an ingenious experiment to determine which of these two types of deficit caused scratching to be a poor instrumental response in rats. A contingency was arranged such that every time a rat scratched itself, two response levers were inserted into the cage, and if the rat then pressed a particular lever (say, the left-hand one) it received a food pellet. When the rat performed some other response (say, rearing) the two levers were again inserted, but this time the rat received a food pellet only if it pressed the other lever (the right-hand one). Thus the rat's own behaviour (scratching or rearing) constituted both an instrumental response reinforced by lever presentation, and a discriminative stimulus telling the rat which lever to press.

Fig. 6.5. (a) Some of the hamster activity patterns which Shettleworth attempted to reinforce.

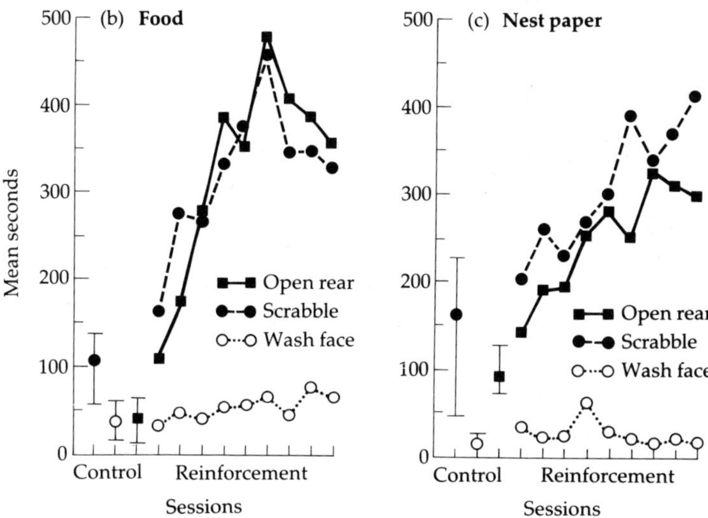

Fig. 6.5. (b) Effect of food reinforcement on scrabbling, rearing and face washing: only scrabbling and rearing increase when reinforced.
(c) Effect of nest-paper reinforcement on the same activities: again only scrabbling and rearing increase when reinforced. (After Shettleworth 1975, 1978.)

Morgan and Nicholas argued that if the rat is unable to form any sort of association between scratching and its consequences, then scratching should be a poor discriminative stimulus as well as a poor instrumental response. On the other hand, if the rat is able to learn an association between scratching and food, but is unable for some other reason to perform the scratching response at an enhanced rate, then scratching should act as a discriminative stimulus despite not being able to act as an instrumental response. The beauty of this experimental design is that it enables the discriminative and instrumental aspects of the task to be measured independently.

The results showed that scratching was both a poor discriminative stimulus (the rats could not learn which lever to press after scratching) and a poor instrumental response (the rate of scratching failed to increase when scratching was reinforced by lever presentation). Rearing, on the other hand, acted both as a good discriminative stimulus and as a good instrumental response. Morgan and Nicholas concluded that the rat suffers from a basic inability to associate scratching with its consequences: the rat behaves as if the information 'I have just scratched myself' is simply not available for use in a learning task.

Pearce *et al.* (1978), on the other hand, found that scratching *would* act as an instrumental response in the rat if a Velcro collar was attached around the rat's neck, despite the fact that the collar did not produce a higher baseline rate of scratching. They concluded that scratching requires for its occurrence an eliciting stimulus or 'itch', provided in their experiment by the collar. Thus, in contrast to Morgan and Nicholas, they attribute the normal failure of scratching to act as an instrumental response to a performance deficit (absence of an itch) rather than to a learning deficit (failure to perceive the association between scratching and food). According to this account, responses such as rearing and scrabbling make good instrumental responses either because they require no specific eliciting stimulus for their occurrence, or because the requisite eliciting stimuli are always present.

We have considered this example in some detail because it shows how difficult it is to prove that inability of an animal to do well in a learning task really is the result of inability to learn. In a normal learning task, learning and performance are inextricably confounded, because performance of a particular response constititutes the measure of learning. The experiments by Morgan and Nicholas, and by Pearce *et al.*, show that with ingenuity one can attempt to disentangle the various factors that might contribute to a constraint on learning. To a sceptical reader this might appear to be conceptual nit-picking, but in fact this sort of argument is crucial, because it forces us to examine more rigorously what we mean by the term 'learning', and what we need to do to demonstrate the occurrence or non-occurrence of learning.

6.3.5 *Differences between reinforcers*

Responses can be conditioned using a wide variety of reinforcers other than food, water and electric shock (for reviews see Glickman & Schiff 1967; Hogan & Roper 1978). Some of these reinforcers, such as opportunity to attack a rival, to court a mate or to explore a novel environment, have obvious significance to the animal; but reinforcers may also be artificial stimuli, such as onset of a light or buzzer, or they may consist of opportunity to perform an arbitrary response such as running in a wheel. Are all these reinforcers equally powerful in their ability to support learning?

From a practical point of view the answer to this question is clearly 'No'. The reinforcing effect of an arbitrary stimulus such as a light or buzzer is at best marginal, requiring for its demonstration complicated statistical tests on data from a number of animals (e.g. Kish 1966). By contrast, to demonstrate that food acts as a reinforcer for a response such as lever-pressing one needs just one hungry rat, and within ten minutes unequivocal conditioning can be shown. The prospect of food seems to concentrate wonderfully the mind of a hungry rat. This is not just because food is in some sense a more natural stimulus than, say, light onset. Stevenson-Hinde (1973) found that the natural stimulus of conspecific song acted only as a weak reinforcer for instrumental learning in the chaffinch, in the sense that the increase in frequency of performance of an instrumental response was slight at best, and was absent in a substantial proportion of birds (Fig. 6.6).

In interpreting these experiments the question again arises: is the animal's failure to do well in the task due to a learning deficit or to a performance deficit (cf. section 6.3.4)? Roper (1973, 1975) found that presentation of nest material appeared at first sight to be a weak reinforcer for key pressing in mice; but factors known to increase nest-building behaviour, such as pregnancy and low temperature, made nest material as powerful a reinforcer as food. Thus, the initial poor performance of the mice stemmed not from a failure to associate response and reinforcement, but from a lack of interest in the activity of nest-building. Similarly, an arbitrary stimulus such as a flashing light could probably be made a powerful reinforcer by endowing the light with more 'motivational significance' for the animal, for example by pairing the light with electric shock so that the light becomes an aversive stimulus. So far, there seems to be no case in which a reinforcer has been shown definitely to produce poor learning as opposed to poor performance of the learned response.

6.3.6 *Avoidance learning*

Just as an animal can learn an instrumental response in order to obtain a desirable reinforcer such as food, so it can learn a response in order to terminate or avoid an undesirable reinforcer such as a

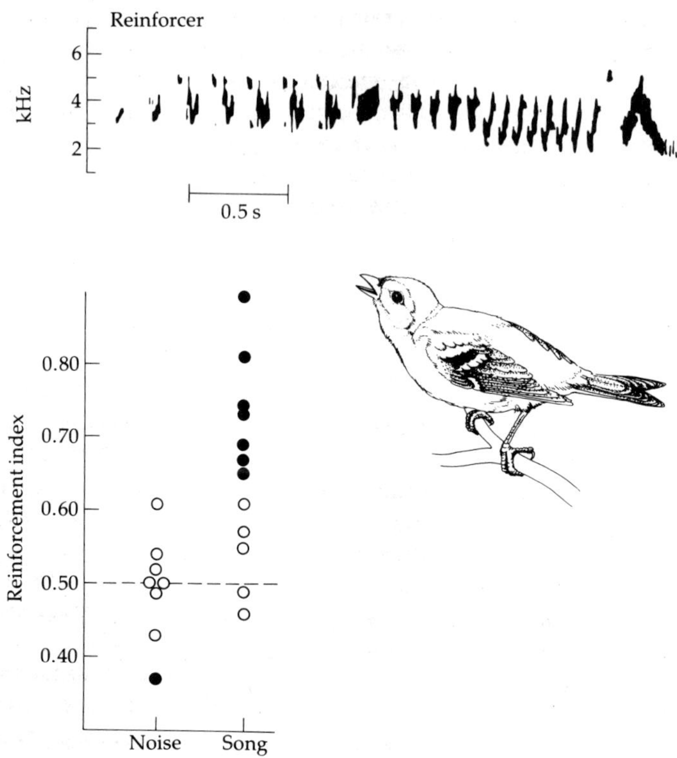

Fig. 6.6. Reinforcement indices for eight individual chaffinches with white noise as reinforcer, and for twelve chaffinches with conspecific song as reinforcer (see sonagram above). The instrumental response was landing on a particular perch in the cage. An index of 0.5 constitutes chance level, and filled circles represent scores differing significantly from chance. Song is a better reinforcer than noise, but even with song the reinforcing effect is very variable, and not all birds learn. (After Stevenson-Hinde 1973.)

loud noise or an electric shock. According to the principle of equipotentiality, any response should be conditionable as an avoidance response, but this appears not to be the case in practice. For example, it is very difficult to train a rat to press a lever, or a pigeon to peck a key, in order to avoid shock. It is also difficult to train a rat to move towards a stimulus that signals shock, but very easy to train a rat to move away from the same stimulus.

From this sort of evidence Bolles (1970, 1978) concluded that avoidance responses can only be conditioned if they are already part of the species' natural defensive behaviour; he referred to such responses as 'species-specific defense reactions'. Bolles's argument is reminiscent of autoshaping (section 6.3.3): responses are more easily conditioned if they resemble behaviour normally elicited by the reinforcer in question. The relevant evidence is reviewed by Bolles (1970), in a paper that constitutes one of the most provocative and cogent of the early contributions to the constraints-on-learning debate.

6.4 Implications for traditional learning theory

6.4.1 *Implications for general process theory*

Critics of the traditional learning theory approach have claimed that the constraints-on-learning evidence challenges general process theory by implying the existence of different mechanisms or processes of learning (e.g. Rozin & Kalat 1971, p. 460; Shettleworth 1972, p. 61). We have already seen (section 6.2.2) that a mechanism or process of learning means, in this context, a learning task that obeys a particular set of laws. Does the constraints-on-learning evidence imply, then, that different learning tasks obey different laws?

In fact, of the 'laws' of learning cited in section 6.2.3 only one is called into question by the constraints-on-learning evidence. This is Law 1, which states that associative learning only occurs when the E1–E2 interval is less than a few seconds long, and it is challenged by food-aversion learning.

As regards the other 'laws', food-aversion learning resembles normal classical conditioning (see reviews by Revusky 1977a; Domjan 1980). For example, conditioning is stronger the more intense the flavour cue; conditioning is stronger the larger the reinforcement (i.e. the more severe the illness); learning extinguishes, though it takes a long time to do so; and it shows the phenomena of overshadowing and blocking. Food-aversion learning is not even qualitatively anomalous with respect to E1–E2 interval because, like other examples of associative learning, it shows a delay-gradient effect (i.e. conditioning gets weaker the longer the E1–E2 interval; see Figs 6.1 and 6.2).

To summarise, the threat posed to general process theory by the constraints-on-learning evidence is weak; only one of the supposed laws of learning is violated, and that only quantitatively. It is not surprising therefore that, rather than abandoning the search for general laws of learning, some psychologists have responded to the evidence by attempting to formulate new 'laws' which can accommodate the phenomenon of food-aversion learning. For example, Revusky (1971, 1977a) proposes that an association is formed not between any E1 and any E2, but between E2 and the most recent *relevant* E1. When E2 is illness, a relevant E1 would be a novel taste, whereas when E2 is a conventional reinforcer such as food or electric shock, any exteroceptive stimulus constitutes a relevant E1. Consequently, an association between a novel taste and illness can survive a long delay so long as no other relevant stimulus (no other novel taste) intervenes (see section 5.3.2 for evidence).

It remains to be seen whether Revusky's new 'law' is useful; obviously it suffers from circularity unless relevance can be defined independently of the phenomena it is supposed to explain. But, before jumping to the conclusion that the search for general laws of learning is still worthwhile, it is worth pointing out that the constraints-on-learning evidence itself uses the methods of the learning theorist: it concerns conditioning in rats and pigeons in laboratory apparatus. Thus the paucity of evidence against general process theory may merely reflect the fact that the theory has scarcely begun to be tested outside a highly restricted methodological domain.

It is also worth pointing out that, although the use of artificial environments and of a limited range of species *may* reveal general principles of learning, psychologists are obviously wrong if they think that it will necessarily do so. The history of theories of motivation provides a good demonstration of this point. It used to be thought that general principles of motivation would emerge from detailed analysis of the feeding and drinking behaviour of the rat in laboratory environments (e.g. Deutsch 1960; Bolles 1967). Subsequent experience has shown, however, that knowledge of how the rat regulates its food and water intake in a laboratory cage does not even begin to tell us how the rat regulates food and water intake in the wild, let alone how it and other species regulate other activities such as sexual behaviour and aggression (e.g. Collier *et al*. 1972; Bolles 1979). By analogy, the only sure way of finding out whether or not laws of learning are general is to test them in as wide a variety of situations as is feasible, including, if possible, natural ones. So far this has not been done to any great extent.

6.4.2 *Implications for the principle of equipotentiality*

Whereas the constraints-on-learning evidence provides very little challenge to general process theory, it appears to undermine to a considerable extent the principle of equipotentiality; the evidence suggests that the ease with which learning occurs varies according to the nature of the stimulus, response and reinforcer, on interactions between these elements of the association, and on the identity of the species involved. It seems reasonable to conclude from this evidence that the differences in learning ability that are observed result from genetically determined differences in the learning mechanism (e.g. Rozin & Kalat 1971). It has subsequently become clear, however, that other interpretations of the evidence are possible.

First, it is important to note that what appears to be a constraint on learning may in fact be a constraint on performance. The reader will recall that in sections 6.3.4 and 6.3.5 we discussed whether apparent differences in the conditionability of different responses, or in the strength of different reinforcers, really resulted from differences in ease of formation of the relevant associations, or from other factors that might loosely be termed 'motivational' or 'attentional' in nature. Similar arguments are currently raging about what precisely underlies the phenomenon of flavour–illness specificity in food-aversion learning (see, for example, Gillette *et al.* 1980; Miller & Domjan 1981). It is too early to predict what the outcome of these arguments will be, but it is fair to say that at the time of writing no constraint on learning has been shown unequivocally to result from a constraint on the animal's ability to acquire information, rather than on its ability to perform a response.

Secondly, some theorists, while assuming that apparent constraints on learning really are caused by constraints on the associative process, have attempted to rescue the principle of equipotentiality in some other way. The most plausible of this second class of counter-arguments relies on the phenomenon of blocking (section 6.2.3), and is most closely associated with Mackintosh (1974).

The essence of Mackintosh's argument is that learning is constrained in adult animals (and all the experiments cited in section 6.3 did involve adults) because of their previous experience, rather than because of intrinsic differences in learning capacity. Blocking is crucial to this idea because it demonstrates that ability to learn can be influenced by previous history of learning. For example, Mackintosh suggests that adult rats are predisposed to associate taste with illness, not because this type of association is somehow innately favoured, but because adult rats have had a lifetime's experience of associating tastes with gastric consequences. In other words, they have learned that after tasting food they experience a particular class of sensations concerned with digestion, which differs from the class of sensations that follows, say, grooming or running around. Consequently, in a food-aversion conditioning experiment, cues such as colour or location of a stimulus, which have not previously been associated with gastric sensations, are blocked by cues such as flavour and odour. A precisely similar argument could be applied to the inability of animals to learn, for instance, that the instrumental response of self-licking produces food reinforcement. In its previous life history the animal would have learned that externally directed responses such as moving around or digging brought it into contact with food, whereas self-directed responses such as grooming did not.

Mackintosh's argument enables the *tabula rasa* doctrine of behaviorism to remain virtually intact: the animal's mind is initially a blank slate, but as the slate is written upon by experience, so the scope for further modification by experience lessens. Put colloquially, one cannot teach an old dog new tricks because during its lifetime the dog accumulates evidence from which it forms particular hypotheses about the way the world is structured. The more evidence it has accumulated favouring one set of hypotheses, the less willing it will be to change these hypotheses in the light of new, conflicting, evidence.

Such an argument seems intuitively to contain at least a grain of truth, but it seems unlikely that the learning process is completely unguided even in very young animals. For example, a predisposition to associate flavour with illness has recently been demonstrated in rat pups and chicks only a few days old (Domjan 1980; Gillette *et al.* 1980), and it is difficult to see what opportunity

LoLordo's pigeons, in their isolated cages, could have had for learning to associate auditory stimuli, rather than visual stimuli, with pain (cf. section 6.3.2).

In conclusion, it can no longer be denied that the performance of adult animals in learning tasks is subject to various kinds of limitation and predisposition. The precise nature and source of these effects, whether or not they really involve learning and whether or not they are themselves a product of learning, remains controversial.

6.5 The ethological approach: adaptive aspects of learning

Ethologists have traditionally been interested in the functional relevance of learning more than in its underlying mechanism. In this respect the ethological and psychological approaches to learning differ at a fundamental level. Nevertheless, evidence from ethology is indirectly relevant to general process theory and directly relevant to the principle of equipotentiality, as well as being of great interest in its own right.

The best-studied examples of natural learning are the acquisition of song in birds and the process of maternal attachment, known as imprinting, which occurs in precocial young of various species. It is therefore inevitable that song learning and imprinting will dominate any discussion of the ethological contribution to knowledge of learning. Song learning is also discussed in section 3.4, and imprinting in sections 2.8 and 4.2.

6.5.1 *Implications for general process theory*

General process theory implies that there is a single learning process which is general in the sense that the same process is applicable in a wide variety of contexts. By contrast, the examples of learning described by ethologists often seem strikingly specific in the following two respects.

First, a single outstanding learning ability is often present in a species that would not otherwise be considered especially intelligent. Consider, for example, the learning of hundreds or even thousands of complex song types in individual birds such as the marsh wren (*Cistothorus palustris*) or the brown thrasher (*Toxostoma rufum*) (Kroodsma 1978; Fig. 6.7a); the extraordinary

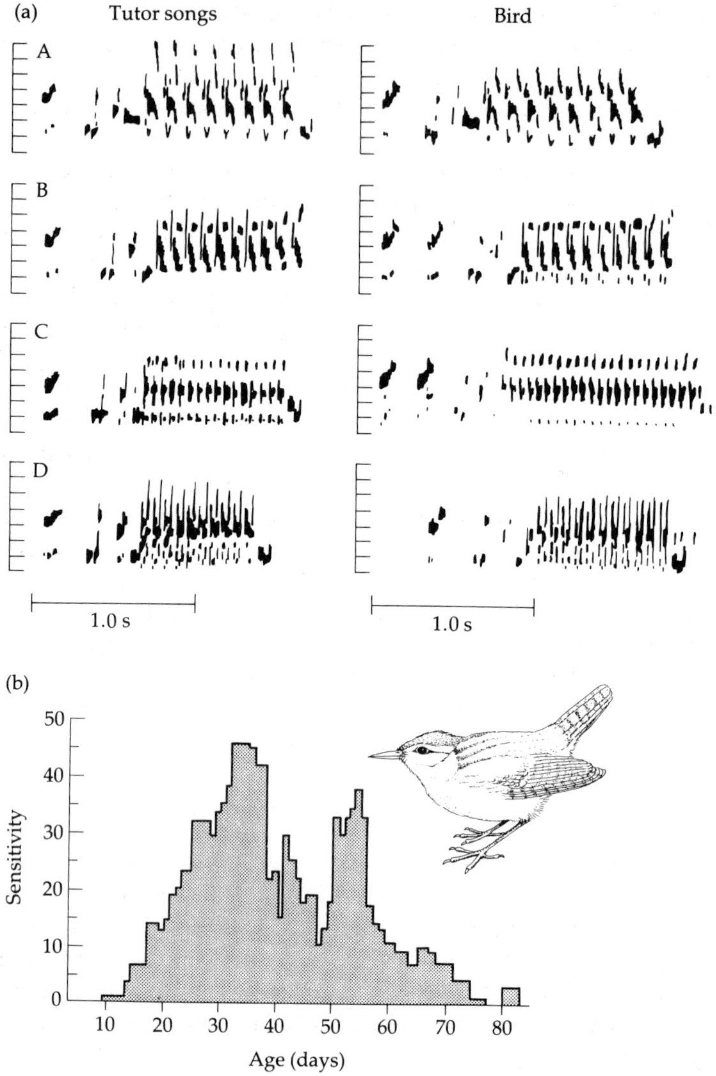

Fig. 6.7. (a) A sample of songs learned from tape-recordings by a single marsh wren. (b) The sensitive period for song learning in the marsh wren. Nine juvenile male birds were exposed to 44 song types, each song type being presented only for a few consecutive days. Since different song types were presented at different ages, the frequency of learning of any one song type constitutes a measure of the sensitivity of the learning process at that particular age. (After Kroodsma 1978.)

rapidity and durability of the learning shown by chicks during imprinting (Bateson 1971); the ability of bees to learn the spatial and temporal distribution of various sources of nectar within the vicinity of the hive (Heinrich 1979); the ability of some fish to recognise their home river when returning from the ocean to spawn (Hasler 1960); and the process whereby indigo buntings develop a navigational reference, for use in nocturnal migrations, by observing the rotation of the night sky around a fixed axis (Emlen 1970). All these feats of learning seem remarkable precisely because they involve species that would be considered unintelligent by comparison with, say, a chimpanzee or a human.

Secondly, in such cases it is often easy to see the specific survival value of that particular learning ability to individuals of the species in question. For example, song learning is useful to birds because song is used to repel rivals and/or attract a mate, and only by learning can the bird be sure to match the song of a neighbour wherever it settles (Krebs & Kroodsma 1980); imprinting is useful to chicks because, being precocial animals, they are likely to wander away from their mother, and imprinting enables them to recognise her from a distance; knowledge of the location of the best source of nectar at a particular time of day is useful to bees because it enables them to forage efficiently.

Because of this corresponding specificity between the incidence of learning and its apparent function, observations of natural learning tend to encourage the view that learning consists, not of a unitary general capacity, but of *a collection of specialised abilities which have evolved independently in particular species in order to do specific jobs*. This latter view does not necessarily contradict general process theory because, even if they have evolved independently, different learning abilities could still conform to a single set of learning laws. However, it seems more likely that learning abilities which have evolved to cope with different tasks would obey different laws (Rozin & Kalat 1971).

6.5.2 *Implications for the principle of equipotentiality*

Whereas the principle of equipotentiality suggests that animals are able to learn almost anything within the limits of their sensory and motor capability, there is direct evidence from ethology of both species- and age-specific limitations on what can be learned.

As regards species-specific limitations, most birds do not copy any song to which they are exposed, but copy only songs which have particular features in common with the songs of their own species (see section 3.4.2). Similarly, a chick does not imprint on any stimulus, it imprints only on stimuli that are within a certain range as regards size and conspicuousness, and in some cases only on a stimulus that emits a species-specific call (Gottlieb 1965; Bateson 1973). An inability to learn which seems to be particularly arbitrary is demonstrated by the herring gull's inability to recognise its own eggs, despite the fact that (a) the eggs show enough variation in colour pattern to be discriminated and (b) herring gulls do learn to recognise their own chicks (Tinbergen 1951). However, such an ability would be redundant, provided that herring gulls can learn the location of their nest, because eggs do not move.

As regards age-specific limitations, song learning in many species of bird can occur only during a particular 'sensitive period' early in the bird's life (Fig. 6.7b; see also section 3.4.1). Similarly, imprinting can only occur when the bird is less than a few days old (Bateson 1971, 1973).

Thus, far from endowing animals with an indiscriminate tendency to acquire information from the environment, evolution seems to have provided an 'innate school-marm' who tells each species what, and when, to learn (Lorenz 1965). Again, it is often easy to see the adaptive value of these restrictions on the learning process; for example, by learning only the song of its own species a bird is assured that its song will attract a conspecific mate, and by having a predisposed preference for certain types of imprinting stimulus the chick increases the chance that it will imprint on its mother, rather than on some inappropriate object such as a stone or another species.

6.5.3 *Variability in the extent to which learning is specific*

The previous examples emphasise the specificity of learning; they imply that discrete learning abilities have evolved in response to discrete selective pressure. However, scientific evidence and common experience show that ability to learn is not always highly specific. If it were, animals would not be able to solve the bizarre problems that psychologists devise for them, and humans would not be able to learn, say, to drive cars. Clearly the ability to learn, once evolved, can often be transferred to novel tasks.

Once again song learning illustrates the variability that can exist even within a single type of learning. As already noted, many birds copy only the songs of their own species or artificially created songs that have the right acoustical characteristics. Other species, however, will copy a wide variety of songs. Similarly, there is bewildering variation in the extent to which song-learning is restricted to a sensitive period (see section 3.4).

At present we have little idea why natural selection has left the song learning process so heavily constrained in some species and so flexible in others. Still less have we any idea why it is possible to teach a monkey to solve a conditional oddity discrimination or a parrot to ride a miniature bicycle, only possible within limits to teach a chimpanzee to use sign language, and completely impossible to teach a horse higher mathematics. These examples do, however, show that it is an oversimplification to ask the question: 'is learning a single general ability or a collection of specific abilities?' It is already evident that learning is less task-specific in some cases than in others; what we need to know is why, and how.

6.6 What is learning for?

As already noted, it is often easy to see why ability to learn certain things is advantageous to members of a particular species. It is also easy to see that certain kinds of information, if they are to be possessed at all, *must* be acquired through learning because they are peculiar to an individual rather than to a species, and hence cannot be known at the time that information is encoded in the genome. For example, an animal can be genetically endowed with information which would allow it to distinguish a conspecific from a heterospecific, because the species' gross morphology and behaviour do not change from generation to generation. But the same animal cannot possess detailed foreknowledge of the appearance of its own mother or its own future mate. Similarly, no animal can be genetically informed of the precise geography of its own territory, which is why one thing that almost every animal must learn is how to find its way around, a kind of learning about which we know very little. But what might be the function of a general associative learning ability of the type postulated by psychologists?

6.6.1 *Perceiving causal relationships*

One suggestion is that associative learning enables the animal to distinguish causal relationships from chance correlations in the environment, and hence to predict the occurrence of biologically significant events (Dickinson 1980; see also Chapter 5). This idea stems from recent interest in phenomena such as blocking, in which the animal seems to learn an association between E1 and E2 to the extent to which E1 predicts E2. In a simple blocking experiment, stimulus A is paired with reinforcement, and then stimulus A plus B is paired with reinforcement. The result is that no association is learned between B and reinforcement. This failure to learn is explained by saying that B provides only redundant information about the occurrence of reinforcement, reinforcement being already perfectly predicted by A.

This view of the function of learning is compatible with the literature on constraints on learning, because it can be argued that if learning has evolved to enable the animal to discern causal relationships, then the properties of learning should mirror those of the kinds of causal relationships that occur in the real world. For example, in the real world causes usually manifest themselves before effects, and correspondingly we find that conditioning usually occurs only when E1 precedes E2. In the real world certain classes of effects result from certain classes of causes (e.g. gastrointestinal disorders usually result from ingestion of something harmful) and, correspondingly, the learning process seems predisposed in favour of these predictable cause–effect relationships.

It remains to be seen whether this way of viewing the learning process can make any genuine predictions about the properties of learning, as opposed to making sense of facts that are already known. Also, to a biologist the 'causal relationships' idea may seem uncomfortably abstract: one can see in a general way that ability to learn about the 'causal texture of the environment' (Dickinson 1980) might be useful to an animal, but how does one translate this into concrete functions in the context of day-to-day behaviour such as food-finding, mate attraction, etc.?

6.2.2 *Solution of technological problems*

An obvious and somewhat more concrete function for instrumental learning is that it enables the animal to solve 'technological' problems encountered, for instance, in dealing with a new source of food. There are certainly some good examples of learned techniques in food-handling, such as termite fishing in chimpanzees, potato washing in macaques, nut opening in squirrels, and milk-bottle opening in blue tits (see review by Galef (1976)). It has even been suggested that chicks learn to drink water by trial-and-error modification of a pecking movement used in eating (Shettleworth 1972). But clear examples of this type of practical problem-solving are rare, given the widespread occurrence of instrumental-learning ability.

6.6.3 *Perceptual sharpening*

Hailman (1969) noted that a very young gull chick will peck at any thin, vertically oriented, moving stimulus, whereas an older chick will peck only at its parent's bill. Similarly, a human baby will initially smile and gurgle at any human face, but later will restrict these responses to its mother's face, and will show fear or indifference towards other faces. A chick which has not been imprinted will approach and twitter at any large, conspicuous, moving object, whereas during imprinting these responses become restricted to one specific object. In all these cases, the range of stimuli which will elicit a response becomes restricted, a process termed 'perceptual sharpening' by Hailman.

Hailman proposed that this process, which is probably of widespread occurrence in all sorts of contexts (e.g. in recognition of prey, avoidance of predators, etc.), is a type of classical conditioning. A similar interpretation of imprinting has been proposed by Hoffman and Ratner (1973; Hoffman 1978). The essence of this idea is that those characteristics of the stimulus which initially elicit the response in question constitute event E2, while other characteristics of the stimulus, which subsequently come to control the response, constitute event E1. Since E1 and E2 are features of the same stimulus complex they are inevitably paired whenever the stimulus is presented.

Consider, for example, a naive chick being imprinted on a

moving green sphere. The stimulus element 'motion' constitutes E2: it elicits filial behaviour in a naive chick. The stimulus elements 'green colour' and 'spherical shape' constitute event E1: they come to elicit filial behaviour during the process of imprinting, by virtue of being paired with E2. Similarly, during the process of facial recognition by a human infant, the stimulus elements which constitute E2 relate to the rough size and shape of a human head, and the presence of eyes; the stimulus elements which constitute E1 are those which specify a particular human head, such as the colour of eyes and hair, the shape of the eyebrows, and so on.

As an interpretation of imprinting, the classical conditioning hypothesis is still controversial (Eiserer 1980; Shettleworth, in press). The main practical problem is that it is difficult to isolate the stimulus elements that are supposed to elicit filial behaviour in the naive chick. There is also a theoretical problem, namely that in imprinting or perceptual sharpening, events E1 and E2 occur simultaneously (both being attributes of a single stimulus complex), an arrangement that would not be expected to lead to strong classical conditioning (see Fig. 6.1). Nevertheless, the process of increasingly sophisticated object recognition which is represented by perceptual sharpening must involve some kind of learning, and because of its probable widespread occurrence it deserves further analysis.

6.6.4 *Learning in a social context*

In contrast to the idea that learning enables the animal to solve practical problems (section 6.6.2), Jolly (1966) and Humphrey (1976) have suggested that learning (or, to be more exact, intellect) is most useful in a social context. More specifically, learning enables the animal to make predictions about the behaviour of conspecifics during social interactions. As Humphrey notes, when an animal is 'playing the game of social plot and counter-plot' it needs all its wits about it; it needs to be quick-thinking in its application of learnt information about the personality characteristics of other members of its social group.

Humphrey's idea has intuitive appeal, but it is difficult to see how it could be tested in the absence of a rigorous definition of intelligence. Here, perhaps, we come up against one of the most fundamental weaknesses in the whole field of learning: our definitions of classical and instrumental conditioning, of intelligence, and of learning itself, are entirely operational in nature. Consequently, there is no guarantee that they reflect distinctions in the underlying mechanism of learning, or that they constitute a useful language with which to discuss the evolution and function of learning.

6.6.5 *The future*

It may strike the reader as curious that so little is known about what learning is, and what it is for, either at the level of a particular species or in general. Ability to learn must, after all, have played a dominant part in determining the success of our own species, while at the same time inability to learn (or at least to act on historical precedent) underlies some of our most tragic shortcomings.

Probably the question 'what is learning for?' has tended not to be asked because it seems self-evident that learning is useful. Now that we can see that the question really is worth asking, the search for concrete answers constitutes one of the most exciting challenges in animal behaviour, and probably also one of the most difficult.

6.7 Selected reading

Biological Boundaries of Learning, edited by Seligman and Hager (1972), contains a useful selection of reprints of early constraints-on-learning papers. It includes the early papers by Garcia and others on food-aversion learning, papers by Thorndike and Konorski on unconditionable instrumental responses, by Jenkins and others on autoshaping, by Bolles and others on avoidance learning, and a selection of material about more natural examples of learning, such as imprinting, bird navigation and language learning in chimps and humans. In addition, the editors provide valuable introductions to the various topics, in which they summarise the historical background and underline the main theoretical issues. This is not a book to read from cover to cover, but an excellent reference book for source material.

Constraints on Learning, edited by Hinde and Stevenson-Hinde (1973), contains useful chapters by Hinde (a general introductory chapter), Bateson (on imprinting), Moore (on autoshaping) and Stevenson-Hinde (on instrumental conditioning).

There have been many reviews of the constraints-on-learning literature, of which the best are by Shettleworth (1972) and Seligman (1970). Good reviews of the food-aversion literature are provided by Rozin and Kalat (1971) and by Domjan (1980). The latter focuses specifically on the question of whether food-aversion learning really is in any sense an exceptional form of learning. Bolles (1970) provides an outstanding review of avoidance learning. Some idea of the current state of the debate about constraints on learning is given by Johnston's (1981) review and by the comments of other interested parties, printed immediately following it.

Kroodsma (1978) gives an excellently clear summary of the relevant facts about bird song learning.

The constraints-on-learning debate is still very much alive, and anyone who wants the very latest points of view should consult current journals such as *Animal Learning and Behaviour; Journal of Experimental Psychology: Animal Behavior Processes; Journal of Comparative and Physiological Psychology*; and *Learning and Motivation.*

REFERENCES

The section(s) in which each reference appears is given after the reference

Aceves-Piña E.O. & Quinn W.G. (1979) Learning in normal and mutant *Drosophila* larvae. *Science* **206,** 93–96.
1.2.4

Alcock J. (1979) *Animal Behavior: An Evolutionary Approach,* 2nd edn. Sinauer, Sunderland, Mass.
2.2.2

Alexander B.K. & Harlow H.F. (1965) Social behaviour of juvenile rhesus monkeys subjected to different rearing conditions during the first six months of life. *Zoologisches Jahrbuch Abteilung. allgemeine Zoologie und Physiologie der Tiere* **71,** 489–508.
4.3.3

Altmann J. (1980) *Baboon Mothers and Infants.* Harvard University Press, Cambridge, Mass.
4.2, 4.2.2, 4.3.1, 4.3.2

Arnold S.J. (1978) Some effects of early experience on feeding responses in the common garter snake, *Thamnophis sirtalis. Animal Behaviour* **26,** 455–462.
3.3.2

Arnold S.J. (1981) Behavioral variation in natural populations II. The inheritance of a feeding response in crosses between geographic races of the garter snake, *Thamnophis elegans. Evolution* **35,** 510–515.
1.6.3

Baptista L.F. & Morton M.L. (1981) Interspecific song acquisition by a white-crowned sparrow. *Auk* **98,** 383–385.
3.4.2

Barlow H.B. (1975) Visual experience and cortical development. *Nature* (London) **258,** 199–204.
3.2

Barrett P. & Bateson P. (1978) The development of play in cats. *Behaviour* **66,** 106–120
4.3.3

Bastock M. (1956) A gene mutation which changes a behavior pattern. *Evolution* **10,** 421–439.
1.4.1

Bateson P.P.G. (1964) Effect of similarity between rearing and testing conditions on chicks' following and avoidance responses. *Journal of Comparative and Physiological Psychology* **57,** 100–103.
2.8

Bateson P.P.G. (1966) The characteristics and context of imprinting. *Biological Reviews* **41,** 177–220.
4.2.1

Bateson P.P.G. (1971) Imprinting. In: *Ontogeny of Vertebrate Behavior* (ed. H. Moltz), pp. 369–387, Academic Press, New York.
6.5.1, 6.5.2

Bateson P.P.G. (1973) Internal influences on early learning in birds. In: *Constraints on Learning* (ed. R.A. Hinde & J. Stevenson-Hinde), pp. 101–116. Academic Press, London.
6.5.2

Bateson P.P.G. (1976a) Specificity and the origins of behavior. In: *Advances in the Study of Behavior* **6,** 1–20.
2.2.1, 2.2.2, 2.7, 3.1, 3.6

Bateson P.P.G. (1976b) Rules and reciprocity in behavioural development. In: *Growing Points in Ethology* (ed. P.P.G. Bateson & R.A. Hinde), pp. 401–421. Cambridge University Press, Cambridge.
2.9, 3.1, 3.6

Bateson P.P.G. (1978a) Early experience and sexual preferences. In: *Biological Determinants of Sexual Behaviour* (ed. J.B. Hutchison), pp. 29–53. John Wiley & Sons, Chichester.
2.7, 2.8

Bateson P.P.G. (1978b) Sexual imprinting and optimal outbreeding. *Nature* (London) **273,** 659–660.
4.2.3

Bateson P.P.G. (1979) How do sensitive periods arise and what are they for? *Animal Behaviour* **27,** 470–486.
2.8, 2.11, 3.4.1, 4.2.1, 4.2.3, 4.5

Bateson P.P.G. (1980) Optimal outbreeding and the development of sexual preferences in Japanese quail. *Zeitschrift für Tierpsychologie* **53,** 231–244.
2.3

Bateson P.P.G. (1981) Ontogeny of behaviour. *British Medical Bulletin* **37,** 159–164.
2.2.1

Bateson P.P.G. (1982) Preferences for cousins in Japanese quail. *Nature* (London) **295,** 236–237.
4.2.3

Bateson P.P.G. & Klopfer P.H. (1982) *Perspectives in Ethology, Vol. 5: Ontogeny.* Plenum Press, New York.
2.11

Bateson P.P.G. & Young M. (1981) Separation from the mother and the development of play in cats. *Animal Behaviour* **29,** 173–180.
4.3.3

Batty J. (1978) Plasma levels of testosterone and male sexual behaviour in strains of the house mouse (*Mus musculus*). *Animal Behaviour* **26,** 339–348.
1.4.2

Beard E.B. (1964) Duck brood behavior at the Seney National Wildlife Refuge. *Journal of Wildlife Management* **28,** 492–521.
4.2.1

Bentley D. (1975) Single gene cricket mutations: effects on behavior, sensilla, sensory neurons, and identified interneurons. *Science* **187,** 760–764.
1.2.1

Benzer S. (1971) From gene to behavior. *Journal of the American Medical Association* **218,** 1015–1022.
1.5

Berman C.M. (1980) Early agonistic experience and rank acquisition among free-ranging rhesus monkeys. *International Journal of Primatology* **1,** 153–170.
4.3.1

Bitterman M.E. (1975) The comparative analysis of learning. *Science* **188,** 699–709.
6.3.1

Bitterman M.E. (1976) Flavor aversion studies. *Science* **192,** 266–267.
6.3.1

Blakemore C. (1977) Genetic instructions and developmental plasticity in the kitten's visual cortex. *Philosophical Transactions of the Royal Society* B **278,** 425–434.
1.2.3, 1.5

Blakemore C. & Cooper G.F. (1970) Development of the brain depends on the visual environment. *Nature* (London) **228,** 477–478.
3.2

Boakes R.A. (1979) Interactions between Type I and Type II processes involving positive reinforcement. In: *Mechanisms of Learning and Motivation* (ed. A. Dickinson & R.A. Boakes), pp. 233–268, Erlbaum, Hillsdale, New Jersey.
6.3.3

Boakes R.A. (in press) Behaviourism and the nature-nurture controversy. In: *Animal Models and Human Behaviour* (ed. G. Davey), John Wiley & Sons, Chichester.
6.2.4

Bolles R.C. (1967) *Theory of Motivation,* Harper & Row, New York.
6.4.1

Bolles R.C. (1970) Species-specific defense reactions and avoidance learning. *Psychological Review* **77,** 32–48.
5.3.3, 6.3, 6.3.6, 6.7

Bolles R.C. (1978) The role of stimulus learning in defensive behavior. In: *Cognitive Processes in Animal Behavior* (ed. S.H. Hulse, H. Fowler & W.K. Honig), pp. 89–107. Erlbaum, Hillsdale, New Jersey.
6.3.6

Bolles R.C. (1979) Toy rats and real rats; nonhomeostatic plasticity in drinking. *Behavioral and Brain Sciences* **1,** 103.
6.4.1

Bonner J.T. (1980) *The Evolution of Culture in Animals.* Princeton University Press, Princeton.
3.6

Bower T.G. (1979) *Human Development.* W.H. Freeman, San Francisco.
4.2.1

Bowlby J. (1951) *Maternal Care and Mental Health.* World Health Organization, Geneva.
4.2.4

Breland K. & Breland M. (1961) The misbehavior of organisms. *American Psychologist* **16,** 681–684.
6.3.3

Brenner S. (1974) The genetics of *Caenorhabditis elegans. Genetics* **77,** 71–94.
1.2.2, 1.5

Bronson G.W. (1968) The fear of novelty. *Psychological Bulletin* **69,** 350–358.
3.3.2

Bronstein P.M., Levine M.J. & Marcus M. (1975) A rat's first bite: the non-genetic cross-generational transfer of information. *Journal of Comparative and Physiological Psychology* **89**, 295–298.
3.3.2

Brown P.L. & Jenkins H.M. (1968) Autoshaping of the pigeon's key-peck. *Journal of the Experimental Analysis of Behavior* **11**, 1–8.
6.3.3

Burghardt G.M. (1967) Chemical-cue preferences of inexperienced snakes: comparative aspects. *Science* **157**, 718–721.
3.3.2

Burghardt G.M. (1971) Chemical cue preferences of newborn snakes: influence of prenatal material experience. *Science* **171**, 921–923.
3.3.2

Burnet B & Connolly K.J. (1981) Gene action and the analysis of behaviour. *British Medical Bulletin* **37**, 107–113.
1.1, 1.7

Bush G.L. (1969) Sympatric host race formation and speciation in frugivorous flies of the genus *Rhagoletis* (Diptera: Tephritidae). *Evolution* **23**, 237–251.
1.6.3

Butler R.A. (1953) Discrimination learning by rhesus monkeys to visual exploration motivation. *Journal of Comparative and Physiological Psychology* **46**, 95–98.
6.3.3

Byers D. (1980) A review of the behavior and biochemistry of *dunce*, a mutation of learning in *Drosophila*. In: *Development and Neurobiology of* Drosophila (ed. O. Siddigui, P. Babu, L. Hall & J. Hall), pp. 467–474. Plenum Press, New York.
1.2.4

Byers J.A. (1980) Play partner preference in Siberian ibex, *Capra ibex sibirica*. *Zeitschrift für Tierpsychologie* **53**, 23–40.
4.3.3

Cade W.H. (1979) The evolution of alternative male reproductive strategies in field crickets. In: *Sexual Selection and Reproductive Competition in Insects* (ed. M. Blum & N.A. Blum), pp. 343–379. Academic Press, London.
1.6.1

Cade W.H. (1981) Alternative male strategies: genetic differences in crickets. *Science* **212**, 563–564.
1.6.1

Caro T.M. (1980a) The effects of experience on the predatory patterns of cats. *Behavioral and Neural Biology* **29**, 1–28.
3.3.2

Caro T.M. (1980b) Effects of mother, object play and adult experience on predation in cats. *Behavioral and Neural Biology* **29**, 29–51.
3.3.2

Carter D.E. & Eckerman D.A. (1976) Reply to Zentall and Hogan. *Science* **191**, 409.
5.4.4

Cassidy J. (1979) Half a century on the concepts of innateness and instinct: survey, synthesis and philosophical implications. *Zeitschrift für Tierpsychologie* **50**, 364–386.
2.2.1

Catchpole C.K. (1982) The evolution of bird sounds in relation to mating and spacing behavior. In: *Evolution and Ecology of Avian Vocalizations* (ed.

D.E. Kroodsma & E.H. Miller). Academic Press, New York.
3.4.4

Caviness V.S. & Rakic P. (1978) Mechanisms of cortical development: a view from mutations in mice. *Annual Review of Neurosciences* **1,** 297–326.
1.3.6

Chalfie M. (in press) Microtubule structure in *Caenorhabditis elegans* neurons. *Cold Spring Harbor Symposium* **46.**
1.2.2

Chalfie M. & Sulston J. (1981) Developmental genetics of the mechanosensory neurons of *Caenorhabditis elegans. Developmental Biology* **82,** 358–370.
1.2.2

Chalfie M. & Thomson J.N. (1979) Organization of neural microtubules in the nematode *Caenorhabditis elegans. Journal of Cell Biology* **82,** 278–289
1.2.2

Chalfie M. & Thomson J.N. (in press) Structural and functional diversity in the neuronal microtubules of *Caenorhabditis elegans. Journal of Cell Biology* **93.**
1.2.2

Chalmers N.R. (1979) *Social Behaviour in Primates.* Edward Arnold, London.
4.5

Chalmers N.R. (1980a) Developmental relationships among social, manipulatory, postural and locomotor behaviours in olive baboons, *Papio anubis. Behaviour* **74,** 22–36.
4.2.1, 4.3.3

Chalmers N.R. (1980b) The ontogeny of play in feral olive baboons (*Papio anubis*). *Animal Behaviour* **28,** 570–585.
4.2, 4.3.3

Cherfas J.J. (1977) Visual system activation in the chick: one-trial avoidance learning affected by duration and patterning of light exposure. *Behavioral Biology* **21,** 52–65.
2.7

Cherfas J.J. (1978) Simultaneous colour discrimination in chicks is improved by brief exposure to light. *Animal Behaviour* **26,** 1–5.
2.7

Cherfas J.J. & Scott A. (1981) Impermanent reversal of filial imprinting. *Animal Behaviour* **29,** 301.
2.8

Clarke B.C. (1979) The evolution of genetic diversity. *Proceedings of the Royal Society* B **205,** 453–474.
1.6.1

Coleman S. (1975) Consequences of response-contingent change in unconditioned stimulus intensity upon the rabbit (*Oryctolagus cuniculus*) nictitating membrane response. *Journal of Comparative and Physiological Psychology* **88,** 591–595.
5.3.3

Collier G., Hirsch E. & Hamlin P.E. (1972) The ecological determinants of reinforcement in the rat. *Physiology and Behavior* **9,** 705–716.
6.4.1

Coppinger R.P. (1970) The effect of experience and novelty on avian feeding behavior with reference to the evolution of warning coloration in butterflies. II. Reactions of naive birds to novel insects. *American Naturalist* **104,** 323–334.
3.3.2

Cruze W.W. (1935) Maturation and learning in chicks. *Journal of Comparative Psychology* **19,** 371–409.
2.7

Curio E. (1975) The functional organization of antipredator behaviour in the pied flycatcher: a study of avian visual perception. *Animal Behaviour* **23,** 1–115.
3.3.1

Curio E. (1978) The adaptive significance of avian mobbing. I. Teleonomic hypotheses and predictions. *Zeitschrift für Tierpsychologie* **48,** 175–183.
3.3.1

Davey G. (1981) *Animal Learning and Conditioning*. Macmillan, London.
5.6

Davies N.B. (1982) Behaviour and competition for scarce resources. In: *Current Problems in Sociobiology* (ed. King's College Sociobiology Group), pp. 363–380. Cambridge University Press, Cambridge.
1.7, 2.9

Davis W.J. (1976) Behavioral and neuronal plasticity in mollusks. In: *Simpler Networks and Behavior* (ed. J.C. Fentress), pp. 224–238. Sinauer, Sunderland, Mass.
1.2.4

Dawkins R. (1976) *The Selfish Gene*. Oxford University Press, Oxford.
4.5

Dawkins R. (1982) *The Extended Phenotype*. W.H. Freeman, San Francisco.
1.5, 4.5

Dempster J.P. (1963) The population dynamics of grasshoppers and locusts. *Biological Reviews* **38,** 490–529.
2.6

Deppe U., Schierenberg E., Cole T., Krieg C., Schmitt D., Yoder B. & von Ehrenstein G. (1978) Cell lineages of the embryo of the nematode *Caenorhabditis elegans*. *Proceedings of the National Academy of Sciences, USA* **75,** 376–380.
1.2.2

Deutsch J.A. (1960) *The Structural Basis of Behaviour*. Cambridge University Press, Cambridge.
6.4.1

Dickinson A. (1980) *Contemporary Animal Learning Theory*. Cambridge University Press, Cambridge.
5.3.2, 5.6, 6.2.1, 6.2.2, 6.6.1

Domjan M. (1980) Ingestional aversion learning: unique and general processes. *Advances in the Study of Behavior* **11,** 276–336.
6.3.1, 6.4.1, 6.4.2, 6.7

Dowsett-Lemaire F. (1979) The imitative range of the song of the marsh warbler, *Acrocephalus palustris,* with special reference to imitations of African birds. *Ibis* **121,** 453–468.
3.4.2, 6.5.3

Dudai Y. (1979) Behavioral plasticity in a *Drosophila* mutant, *dunce DB 276*. *Journal of Comparative Physiology* **130,** 271–275.
1.2.4

Dudai Y. & Bicker G. (1978) Comparison of visual and olfactory learning in *Drosophila*. *Naturwissenschaften* **65,** 494–495.
1.2.4

Dudai Y. & Quinn W.G. (1980) Genes and learning in *Drosophila*. *Trends in Neurosciences* February, 28–30.
1.2.4

Dudai Y., Jan Y.–N., Byers D., Quinn W.G. & Benzer S. (1976) *dunce*, a mutant of *Drosophila* deficient in learning. *Proceedings of the National Academy of Sciences, USA* **73,** 1684–1688.
1.2.4, 2.6

Dudley D. (1974) Paternal behavior in the California mouse (*Peromyscus californicus*). *Behavioral Biology* **11,** 247–252.
4.3.2

Dunbar R.I.M. (1982) Intraspecific variations in mating strategy. In: *Perspectives in Ethology, Vol. 5: Ontogeny* (ed. P.P.G. Bateson & P.H. Klopfer). Plenum Press, New York.
2.9

Dunn J.F. (1976) Mother–infant relations: continuities and discontinuities over the first 14 months. *Journal of Psychosomatic Research* **20,** 273–277.
4.2.3

Dusenbery D.B. (1980) Chemotactic behavior of mutants of the nematode *Caenorhabditis elegans* that are defective in osmotic avoidance. *Journal of Comparative Physiology* **137,** 93–96.
1.2.2

Dusenbery D.B., Sheridan R.E. & Russell R.L. (1975) Chemotaxis-defective mutants of the nematode *Caenorhabditis elegans*. *Genetics* **80,** 297–309.
1.2.2

Ehrman L. & Parsons P.A. (1981) *Behavior Genetics and Evolution*. McGraw-Hill, New York.
1.1

Eibl-Eibesfeldt I. (1963) Angeborenes und Erworbenes im verhalten einiger Säuger. *Zeitschrift für Tierpsychologie* **20,** 705–754.
5.3.3

Eibl-Eibesfeldt I. (1970) *Ethology: The Biology of Behavior*. Holt, Rinehart & Winston, New York.
2.2.2

Einon D.F., Morgan M.J. & Kibbler C.C. (1978) Brief periods of socialization and later behaviour in the rat. *Developmental Psychobiology* **11,** 213–225.
4.3.3

Eiserer L.A. (1980) Development of filial attachment to static and visual features of an imprinting object. *Animal Learning and Behavior* **8,** 159–166.
6.6.3

Elwood R.W. (1975) Paternal and maternal behaviour in the Mongolian gerbil. *Animal Behaviour* **23,** 766–772.
4.3.2

Elwood R.W. (1980) The development, inhibition and disinhibition of pup-cannibalism in the Mongolian gerbil. *Animal Behaviour* **28,** 1188–1194.
4.3.2

Elwood R.W. & Broom D.M. (1978) The influence of litter size and parental behaviour on the development of Mongolian gerbil pups. *Animal Behaviour* **26,** 438–454.
4.3.2

Emlen S.T. (1970) Celestial rotation: its importance in the development of migratory orientation. *Science* **170,** 1198–1201.
6.5.1

Estes R.D. & Estes R.R. (1979) The birth and survival of wildebeest calves. *Zeitschrift für Tierpsychologie* **50,** 45–95.
4.2.1

Evans S.M. (1966a) Non-associative avoidance learning in nereid polychaetes. *Animal Behaviour* **14,** 102–106.
5.2.2

Evans S.M. (1966b) Non-associative behavioural modifications in the polychaete *Nereis diversicolor. Animal Behaviour* **14,** 107–119.
5.2.2

Fagen R. (1981) *Animal Play Behaviour.* Oxford University Press, Oxford.
4.3.3, 4.4

Falconer D.S. (1981) *An Introduction to Quantitative Genetics.* Longman, London.
1.4.1, 1.4.2

Feldman M.W. & Lewontin R.C. (1975) The heritability hang-up. *Science* **190,** 1163–1168.
1.5, 2.5

Fisher J. & Hinde R.A. (1949) The opening of milk bottles by birds. *British Birds* **42,** 347–357.
2.3, 3.5

Foree D.D. & LoLordo V.M. (1973) Attention in the pigeon: the differential effects of food-getting vs. shock-avoidance procedures. *Journal of Comparative and Physiological Psychology* **85,** 551–558.
6.3.2

Frisch K. von (1974) Decoding the language of the bee. *Science* **185,** 663–668.
3.5

Fulker D.W. (1966) Mating speed in male *Drosophila melanogaster:* a psychogenetic analysis. *Science* **153,** 203–205.
1.4.4

Fuller J.L. & Thompson W.R. (1978) *Foundations of Behavior Genetics.* Mosby, Saint Louis.
1.1, 1.4.3

Galef B.G. Jr (1976) The social transmission of acquired behavior: a discussion of tradition and social learning in vertebrates. *Advances in the Study of Behavior* **6,** 77–100.
2.3, 3.3.2, 3.6, 6.6.2

Galef B.G. Jr & Sherry D.F. (1973) Mother's milk: a medium for transmission of cues reflecting the flavor of mother's diet. *Journal of Comparative and Physiological Psychology* **83,** 374–378.
3.3.2

Garcia J. & Hankins W.G. (1977) On the origin of food aversion paradigms. In: *Learning Mechanisms in Food Selection* (ed. L.M. Barker, M.R. Best & M. Domjan), pp. 3–19. Baylor University Press, Waco, Texas.
3.3.2

Garcia J. & Koelling R.A. (1966) Relation of cue to consequence in avoidance learning. *Psychonomic Science* **4,** 123–124.
6.3.1, 6.3.2

Garcia J., Ervin F.R. & Koelling R.A. (1966) Learning with prolonged delay of reinforcement. *Psychonomic Science* **5,** 121–122.
3.3.2, 6.3.1

Garcia J.F., McGowan B.K. & Green K.F. (1972) Biological constraints on conditioning. In: *Classical Conditioning II: Current Research and Theory* (ed. A.H. Black & W.F. Prokasy), pp. 3–27. Appleton-Century-Crofts, New York.
6.2.4

Garcia J., Clarke J.C. & Hankins W.G. (1973) Natural responses to scheduled rewards. In: *Perspectives in Ethology* (ed. P.P.G. Bateson & P.H. Klopfer), pp. 1–41. Plenum Press, New York.
6.3

Gaze R.M. & Keating M.J. (1972) The visual system and 'neuronal specificity'. *Nature* (London) **237,** 375–378.
1.2.3

Gerhardt J., Ubbels G., Black S., Hara K. & Kirschner M. (1981) Reinvestigation of the role of the grey crescent in axis formation in *Xenopus laevis. Nature* (London) **293,** 511–516.
1.5

Gillette K., Martin G.M. & Bellingham P. (1980) Differential use of food and water cues in the formation of conditioned aversions by domestic chicks (*Gallus gallus*). *Journal of Experimental Psychology: Animal Behavior Processes* **6,** 99–111.
3.3.2, 6.3.1, 6.4.2

Glickman S.E. & Schiff B.B. (1967) A biological theory of reinforcement. *Psychological Review* **74,** 81–109.
6.3.5

Gottlieb G. (1961) Developmental age as a baseline for determination of the critical period in imprinting. *Journal of Comparative and Physiological Psychology* **54,** 422–427.
2.8

Gottlieb G. (1965) Imprinting in relation to parental and species identification by avian neonates. *Journal of Comparative and Physiological Psychology* **59,** 345–356.
6.5.2

Gottlieb G. (1971) *Development of Species Identification in Birds.* University of Chicago Press, Chicago.
2.2.1

Gottlieb G. (1973) Neglected developmental variables in the study of species identification in birds. *Psychological Bulletin* **79,** 362–372.
2.2.1

Gottlieb G. (1974) On the acoustic basis of species identification in wood ducklings (*Aix sponsa*). *Journal of Comparative and Physiological Psychology* **87,** 1038–1048.
4.2.1

Gottlieb G. (1976a) The roles of experience in the development of behavior and the nervous system. In: *Neural and Behavioral Specificity: Studies in the Development of Behavior and the Nervous System* (ed. G. Gottlieb), pp. 25–54. Academic Press, New York.
2.7

Gottlieb G. (1976b) Early development of species-specific auditory perception in birds. In: *Neural and Behavioral Specificity: Studies in the Development of Behavior and the Nervous System* (ed. G. Gottlieb), pp. 237–280. Academic Press, New York.
2.2.1

Gottlieb G. (1980) Development of species identification in ducklings: VI. Specific embryonic experience required to maintain species-typical perception in Peking ducklings. *Journal of Comparative and Physiological Psychology* **94,** 579–587.
2.2.1

Green R., Carr W.S. & Green M. (1968) The hawk–goose phenomenon: further confirmation and the search for the releaser. *Journal of Psychology* **69,** 271–276.
3.3.1

Grohmann J. (1939) Modifikation oder Funktionsreifung? *Zeitschrift für Tierpsychologie* **2,** 132–144.
5.1

Groves P.M. & Thompson R.F. (1970) Habituation: a dual-process theory. *Psychological Review* **77,** 419–450.
5.2.2

Guillery R.W. (1974) Visual pathways in albinos. *Scientific American* **230** (5), 44–54.
1.2.3

Guillery R.W., Casagrande V.A. & Oberdorfer M.D. (1974) Congenitally abnormal vision in Siamese cats. *Nature* (London) **252,** 195–199
1.2.3

Guiton P. (1959) Socialisation and imprinting in brown leghorn chicks. *Animal Behaviour* **7,** 26–34.
2.8

Gustavson C.R. (1977) Comparative and field aspects of learned food aversions. In: *Learning Mechanisms in Food Selection* (ed. L.M. Barker, M.R. Best & M. Domjan), pp. 23–43. Baylor University Press, Waco, Texas.
3.3.2

Gwinner E. & Wiltschko W. (1980) Circannual changes in migratory orientation of the garden warbler *Sylvia borin. Behavioral Ecology and Sociobiology* **7,** 73–78.
2.3

Hailman J.P. (1967) The ontogeny of an instinct. *Behaviour* Suppl. 15.
2.2.2, 3.2

Hailman J.P. (1969) How an instinct is learnt. *Scientific American* **221** (6), 98–108.
3.6, 6.6.3

Hall J.C. (1977) Portions of the central nervous system controlling reproductive behavior in *Drosophila melanogaster. Behavior Genetics* **7,** 291–312.
1.3.4

Hall J.C. (1979) Control of male reproductive behavior by the central nervous system of *Drosophila:* dissection of a courtship pathway by genetic mosaics. *Genetics* **92,** 437–457.
1.3.4

Hall J.C., Gelbart W.M. & Kankel D.R. (1976) Mosaic systems. In: *The Genetics and Biology of Drosophila, Vol. 1* (ed. M. Ashburner & E. Novitski), pp. 265–314. Academic Press, London.
1.3.2

Handler A.N. & Konopka R.J. (1979) Transplantation of a circadian pacemaker in *Drosophila. Nature* (London) **279,** 236–238.
1.3.3

Harlow H.F. (1949) The formation of learning sets. *Psychological Review* **56,** 51–65.
5.4.2

Harris M.A. & Lemon R.E. (1972) Songs of song sparrows: individual variation and dialects. *Canadian Journal of Zoology* **50,** 301–309.
3.4.3

Hasler A.D. (1960) Guideposts of migrating fishes. *Science* **132,** 785–792.
6.5.1

Hearst E. (1978) Stimulus relationships and feature selection in learning and behavior. In: *Cognitive Processes in Animal Behavior* (ed. S.H. Hulse, H. Fowler & W.K. Honig), pp. 51–88, Erlbaum, Hillsdale, New Jersey.
6.3.3

Hebb D.O. (1953) Heredity and environment in mammalian behaviour. *British Journal of Animal Behaviour* **1,** 43–47.
Introduction

Hedgecock E.M. & Russell R.L. (1975) Normal and mutant thermotaxis in the nematode *Caenorhabditis elegans. Proceedings of the National Academy of Sciences, USA* **72,** 4061–4065.
1.2.2

Heinrich B. (1979) 'Majoring' and 'minoring' by foraging bumblebees, *Bombus vagans:* an experimental analysis. *Ecology* **60,** 245–255.
6.5.1

Herman L.M. & Gordon J.A. (1974) Auditory delayed matching in the bottlenose dolphin. *Journal of the Experimental Analysis of Behavior* **21,** 19–29.
5.4.4

Herman R. & Horvitz R. (1980) Genetic analysis of *Caenorhabditis elegans.* In: *Nematodes as Biological Models* (ed. B. Zuckerman), pp. 227–264. Academic Press, New York.
1.2.2

Hess E.H. (1959) Imprinting. *Science* **130,** 133–141.
4.2.1

Hinde R.A. (1968) Dichotomies in the study of development. In: *Genetic and Environmental Influences on Behaviour* (ed. J.M. Thoday & A.S. Parkes), pp. 3–14. Oliver & Boyd, Edinburgh.
2.4

Hinde R.A. (1974) *Biological Bases of Human Social Behaviour.* McGraw-Hill, New York.
4.2.2

Hinde R.A. (1977) Mother–infant separation and the nature of inter-individual relationships: experiments with rhesus monkeys. *Proceedings of the Royal Society of London* B **196,** 29–50.
4.2.3, 4.2.4

Hinde R.A. (1979) *Towards Understanding Relationships.* Academic Press, London.
4.1, 4.5

Hinde R.A. (1982) *Ethology.* Fontana, London.
2.11

Hinde R.A. & Fisher J. (1951) Further observations on the opening of milk bottles by birds. *British Birds* **44,** 392–396.
2.3, 3.5

Hinde R.A. & Stevenson-Hinde J. (eds) (1973) *Constraints on Learning.* Academic Press, London.
6.3, 6.7

Hinde R.A., Thorpe W.H. & Vince M.A. (1956) The following response of young coots and moorhens. *Behaviour* **9,** 214–242.
2.8

Hirsch H.V.B. & Spinelli D.N. (1970) Visual experience modifies distribution of horizontally and vertically oriented receptive fields in cats. *Science* **168,** 869–871.
3.2

Hirsch J. (ed) (1967) *Behavior–Genetic Analysis.* McGraw-Hill, New York.
1.1

Hoffman H.S. (1978) Experimental analysis of imprinting and its behavioral effects. *The Psychology of Learning and Motivation* **12,** 1–39.
6.6.3

Hoffman H.S. & Ratner A.M. (1973) A reinforcement model of imprinting: implications for socialization in monkeys and men. *Psychological Review* **80,** 527–544.
6.6.3

Hogan J.A. (1973) Development of food recognition in young chicks: II. Learned associations over long delays. *Journal of Comparative and Physiological Psychology* **83,** 367–373.
3.3.2

Hogan J.A. (1975) Development of food recognition in young chicks: III. Discrimination. *Journal of Comparative and Physiological Psychology* **89,** 95–104.
3.3.2

Hogan J.A. & Roper T.J. (1978) A comparison of the properties of different reinforcers. *Advances in the Study of Behavior* **8,** 156–255.
6.3, 6.3.5

Horn G., Rose S.P.R. & Bateson P.P.G. (1973) Experience and plasticity in the central nervous system. *Science* **181,** 506–514.
2.8

Hotta Y. & Benzer S. (1970) Genetic dissection of the *Drosophila* nervous system by means of mosaics. *Proceedings of the National Academy of Sciences, USA* **67,** 1156–1163.
1.3.2

Hotta Y. & Benzer S. (1972) Mapping of behaviour in *Drosophila* mosaics. *Nature* (London) **240,** 527–535.
1.3.2, 1.5

Hotta Y. & Benzer S. (1976) Courtship in *Drosophila* mosaics: sex-specific foci for sequential action patterns. *Proceedings of the National Academy of Sciences, USA* **73,** 4154–4158.
1.3.4

Hubel D.H. & Wiesel T.N. (1963) Receptive fields of cells in striate cortex of very young, visually inexperienced kittens. *Journal of Neurophysiology* **26,** 994–1002.
3.2

Humphrey G. (1933) *The Nature of Learning.* Routledge & Kegan Paul, London.
5.2.1

Humphrey N.K. (1976) The social function of intellect. In: *Growing Points in Ethology* (ed. P.P.G. Bateson & R.A. Hinde), pp. 303–317. Cambridge University Press, Cambridge.
6.6.4

Hutchison R.E. & Bateson P.P.G. (in press) Sexual imprinting in male Japanese quail: the effects of castration at hatching. *Developmental Psychobiology*.
4.2.3

Immelmann K. (1969) Song development in the zebra finch and other estrildid finches. In: *Bird Vocalizations* (ed. R.A. Hinde), pp. 61–74. Cambridge University Press, Cambridge.
3.4.2

Immelmann K. (1972) Sexual and other long-term aspects of imprinting in birds and other species. *Advances in the Study of Behavior* **4,** 147–174.
2.3, 2.4, 2.7, 4.2.1, 4.2.3

Immelmann K. (1975) The evolutionary significance of early experience. In: *Function and Evolution in Behaviour* (ed. G. Baerends, C. Beer & A. Manning), pp. 243–253. Clarendon Press, Oxford.
4.2.1

Immelmann K., Barlow G.W., Petrinovich L. & Main M. (1981) *Behavioral Development: The Bielefeld Interdisciplinary Project.* Cambridge University Press, Cambridge.
2.11

Ince S.A., Slater P.J.B. & Weismann C. (1980) Changes with time in the songs of a population of chaffinches. *Condor* **82,** 285–290.
3.4.3

Jacobs J. (1981) How heritable is innate behaviour? *Zeitschrift für Tierpsychologie* **55,** 1–18.
2.2.1, 2.2.2, 2.5

Jennings H.S. (1906) *Behavior of the Lower Organisms.* Columbia University Press, New York.
5.2.1

Jensen A.R. (1969) How much can we boost IQ and scholastic achievement? *Harvard Educational Review* **39,** 1–123.
Introduction

Johnston T.D. (1981) Contrasting approaches to a theory of learning. *Behavioral and Brain Sciences* **4,** 125–173.
6.7

Johnston T.D. & Gottlieb G. (1981) Development of visual species identification in ducklings: what is the role of imprinting? *Animal Behaviour* **29,** 1082–1109.
4.2.1

Jolly A. (1966) Lemur social behavior and primate intelligence. *Science* **153,** 501–506.
6.6.4

Kandel E.R. (1976) *Cellular Basis of Behavior.* W.H. Freeman, San Francisco.
5.2.1, 5.6

Kankel D.R. & Hall J.C. (1976) Fate mapping of nervous system and other internal tissues in genetic mosaics of *Drosophila melanogaster. Developmental Biology* **48,** 1–24.
1.3.2

Kear J. (1962) Food selection in finches with special reference to interspecific differences. *Proceedings of the Zoological Society of London* **138,** 163–204.
3.3.2

Kear J. (1967) Experiments with young nidifugous birds on a visual cliff. *Wildfowl Trust Annual Report* **18,** 122–124.
3.2

Keating M.J. (1974) Visual function and binocular visual connexions. *British Medical Bulletin* **30,** 145–151.
3.2

Kimble G.A. (1961) *Hilgard and Marquis' Conditioning and Learning,* Appleton-Century-Crofts, New York.
5.1, 6.2.3, 6.3.5

Kimble J.E. (1981a) Strategies for control of pattern formation in *Caenorhabditis elegans. Philosophical Transactions of the Royal Society of London* B **295,** 539–551.
1.2.2

Kimble J.E. (1981b) Alterations in cell lineage following laser ablation of cells in the somatic gonad of *Caenorhabditis elegans. Developmental Biology* **87,** 286–300.
1.2.2

Kimble J.E. & Hirsh D. (1979) The post embryonic cell lineages of the hermaphrodite and male gonads in *Caenorhabditis elegans. Developmental Biology* **70,** 369–417.
1.2.2

Kimble J.E. & White J.G. (1981) On the control of germ cell development in *Caenorhabditis elegans. Developmental Biology* **81,** 208–219.
1.2.2

Kish G.B. (1966) Studies of sensory reinforcement. In: *Operant Behavior: Areas of Research and Application* (ed. W.K. Honig), pp. 10–159. Appleton-Century-Crofts, New York.
6.3.5

Klein M. & Kandel E.F. (1978) Presynpatic modulation of voltage-dependent Ca^{2+} current: mechanism for behavioral sensitization in *Aplysia californica. Proceedings of the National Academy of Sciences, USA* **75,** 3512–3516.
1.2.4

Klopfer P. & Klopfer M. (1977) Compensatory responses of goat mothers to their impaired young. *Animal Behaviour* **25,** 286–291.
2.9

Köhler W. (1925) *The Mentality of Apes.* Routledge & Kegan Paul, London.
5.4.1

Konopka R.J. & Benzer S. (1971) Clock mutants of *Drosophila melanogaster. Proceedings of the National Academy of Sciences, USA* **68,** 2112–2116.
1.3.3

Konopka R.J. & Wells S. (1980) *Drosophila* clock mutations affect the morphology of a brain neurosecretory cell group. *Journal of Neurobiology* **11,** 411–415.
1.3.3

Konorski J. (1967) *Integrative Activity of the Brain.* University of Chicago Press, Chicago.
6.3.4

Krebs J.R. & Davies N.B. (1981) *An Introduction to Behavioural Ecology.* Blackwell Scientific Publications, Oxford.
1.6.1

Krebs J.R. & Kroodsma D.E. (1980) Repertoires and geographical variation in bird song. *Advances in the Study of Behaviour* **11,** 143–177.
6.5.1

Kroodsma D.E. (1978) Aspects of learning in the ontogeny of bird song: where, from whom, when, how many, which and how accurately? In: *The*

Development of Behavior: Comparative and Evolutionary Aspects (ed. G. Burghardt & M. Bekoff), pp. 215–230. Garland, New York.
3.4.1, 6.5.1, 6.5.3, 6.7

Kroodsma D.E. (1980) Winter wren singing behavior: a pinnacle of song complexity. *Condor* **82,** 357–365.
3.4.3

Kroodsma D.E. & Pickert R. (1980) Environmentally dependent sensitive periods for avian vocal learning. *Nature* (London) **288,** 477–479.
3.4.1

Kyriacou C.P. & Hall J.C. (1980) Circadian rhythm mutations in *Drosophila melanogaster* affect short-term fluctuations in the male's courtship song. *Proceedings of the National Academy of Sciences, USA* **77,** 6729–6733.
1.3.3

Landsberg J.-W. (1976) Posthatch age and developmental age as a baseline for determination of the sensitive period for imprinting. *Journal of Comparative and Physiological Psychology* **90,** 47–52.
2.8

Lawick-Goodall J. van (1968) The behaviour of free-living chimpanzees in the Gombe Stream Reserve. *Animal Behaviour Monographs* **1,** 165–311.
4.2.1

Lawrence P.A. (1981) A general cell marker for clonal analysis of *Drosophila* development. *Journal of Embyology and Experimental Morphology* **64,** 321–332.
1.3.2

Lehrman D.S. (1970) Semantic and conceptual issues in the nature–nurture problem. In: *Development and Evolution of Behavior* (ed. L.R. Aronson, E. Tobach, D.S. Lehrman & J.S. Rosenblatt), pp. 17–52. W.H. Freeman, San Francisco.
Introduction, 1.1, 1.5, 2.2.1, 2.2.2, 2.11

Lerner I.M. (1954) *Genetic Homeostasis.* Oliver & Boyd, Edinburgh.
1.4.2

Lett B.T. (1980) Taste potentiates color-sickness associations in pigeons and quail. *Animal Learning and Behavior* **8,** 193–198.
6.3.1

Levitsky D.A. (1979) Malnutrition and the hunger to learn. In: *Malnutrition, Environment and Behavior* (ed. D.A. Levitsky), pp. 161–179. Cornell University Press, Ithaca.
2.9

Lewis J.A. & Hodgkin J.A. (1977) Specific neuroanatomical changes in chemosensory mutants of the nematode *Caenorhabditis elegans. Journal of Comparative Neurology* **172,** 489–510.
1.2.2

Livingstone M.S. (1981) Two mutations in *Drosophila* affect the synthesis of octopamine, dopamine and serotonin by altering the activities of two different amino acid decarboxylases. *Neuroscience Abstracts* **7,** 351.
1.3.3

Locke-Haydon J. & Chalmers N.R. (1983) The development of infant–caregiver relationships in captive common marmosets (*Callithrix jacchus*). *International Journal of Primatology* **4.**
4.3.2

LoLordo V.M. (1979) Selective associations. In: *Mechanisms of Learning and Motivation* (ed. A. Dickinson & R.A. Boakes), pp. 367–398. Erlbaum, Hillsdale, New Jersey.
6.3.2

Lorenz K. (1939) Vergleichende Verhaltensforschung. *Zoologischer Anzeiger Supplement* **12,** 69–102.
3.3.1

Lorenz K. (1965) *Evolution and Modification of Behavior.* University of Chicago Press, Chicago.
Introduction, 2.2.1, 2.3, 6.5.2

Lund R.D. (1978) *Development and Plasticity of the Brain.* Oxford University Press, New York.
1.2.3

Macagno E.R. (1980) Genetic approaches to invertebrate neurogenesis. *Current Topics in Developmental Biology* **15,** 319–345.
1.7

McCance R.A. (1962) Food, growth, and time. *Lancet* **2,** 671–676.
2.9, 3.1

McFarland D.J. (1977) Decision making in animals. *Nature* (London) **269,** 15–21.
3.5

Mackintosh N.J. (1974) *The Psychology of Animal Learning.* Academic Press, London.
5.6, 6.2.3, 6.4.2

Mackintosh N.J. (1975) A theory of attention: variations in the associability of stimuli with reinforcement. *Psychological Review* **82,** 276–298.
5.3.2

Mackintosh N.J. & Dickinson A. (1979) Instrumental (Type II) conditioning. In: *Mechanisms of Learning and Motivation* (ed. A. Dickinson & R.A. Boakes), pp. 143–169. Erlbaum, Hillsdale, New Jersey.
5.3.2

Manning A. (1975) The place of genetics in the study of behaviour. In: *Growing Points in Ethology* (ed. P.P.G. Bateson & R.A. Hinde), pp. 327–343. Cambridge University Press, Cambridge.
1.1, 1.7

Manning A. (1976) Behaviour genetics and the study of behavioural evolution. In: *Function and Evolution in Behaviour.* (ed. G. Baerends, C. Beer & A. Manning), pp. 71–91. Oxford University Press, Oxford.
1.1, 1.7

Manning A. (1979) *An Introduction to Animal Behaviour.* Edward Arnold, London.
5.6

Marler P. (1970) A comparative approach to vocal learning: song development in white-crowned sparrows. *Journal of Comparative and Physiological Psychology* **71** (Supplement), 1–25.
3.4

Marler P. (1976) Sensory templates in species-specific behavior. In: *Simpler Networks and Behavior* (ed. J.C. Fentress), pp. 314–329. Sinauer, Sunderland, Mass.
2.2.2, 3.4

Marler P. (1981) Birdsong: the acquisition of a learned motor skill. *Trends in Neurosciences* **3,** 88–94.
3.4, 3.4.3, 3.6, 6.5.2

Marler P. & Mundinger P.C. (1975) Vocalisations, social organisation and breeding biology of the twite *Acanthus flavirostris. Ibis* **117,** 1–17.
3.4

Marler P. & Peters S. (1977) Selective vocal learning in a sparrow. *Science* **198,** 519–521.
3.4.2, 6.5.2

Marler P. & Peters S. (1981) Sparrows learn song and more from memory. *Science* **213,** 780–782.
3.4, 3.4.2

Marlin N.A. & Miller R.R. (1981) Associations to contextual stimuli as a determinant of long-term habituation. *Journal of Experimental Psychology: Animal Behavior Processes* **7,** 313–333.
5.2.1

Mather K. & Jinks J.L. (1971) *Biometrical Genetics*. Chapman & Hall, London.
1.4.1, 1.4.4

Medioni J. & Vaysse G. (1975) Suppression conditionelle d'un reflexe chez la Drosophile (*Drosophila melanogaster*): acquisition et extinction. *Comptes Rendus Societé Biologique* **169,** 1386.
1.2.4

Menne D. & Spatz H.C. (1977) Color learning in *Drosophila. Journal of Comparative Physiology* **114,** 301–312.
1.2.4

Miller V. & Domjan M. (1981) Specificity of cue to consequence in aversion learning in the rat: control for US-induced differential orientations. *Animal Learning and Behaviour* **9,** 339–345.
6.4.2

Mineka S. & Suomi S.J. (1978) Social separation in monkeys. *Psychological Bulletin* **85,** 1376–1400.
4.2.4

Mishkin M. & Delacour J. (1975) An analysis of short-term visual memory in the monkey. *Journal of Experimental Psychology: Animal Behavior Processes* **1,** 326–334.
5.4.4

Mitchell D.E. (1978) Effect of early visual experience on the development of certain perceptual abilities in animals and man. In: *Perception and Experience* (ed. R.D. Walk & H.L. Pick Jr), pp. 37–75. Plenum Press, New York.
3.2

Miyadi D. (1964) Social life of Japanese monkeys. *Science* **143,** 783–786.
3.5

Moore B.R. (1973) The role of directed Pavlovian reactions in simple instrumental learning in the pigeon. In: *Constraints on Learning: Limitations and Predispositions* (ed. R.A. Hinde & J. Stevenson-Hinde), pp. 159–188. Academic Press, London.
6.3.3

Morgan M.J. & Nicholas D.J. (1979) Discrimination between reinforced action patterns in the rat. *Learning and Motivation* **10,** 1–22.
6.3.4

Morris R.G.M. (1981) Spatial localization does not require the presence of local cues. *Learning and Motivation* **12,** 239–260.
5.4.3

Mueller H.C. & Parker P.G. (1980) Naive ducklings show different cardiac responses to hawk than to goose models. *Behaviour* **74,** 101–113.
3.3.1

Mullen R.J. (1978) Mosaicism in the central nervous system of mouse chimeras. In: *Clonal Basis of Development* (ed. S. Subtelny & I.M. Sussex). Academic Press, New York.
1.3.5

Mullen R.J. & Herrup K (1979) Chimeric analysis of mouse cerebellar mutants. In: *Neurogenetics: Genetic Approaches to the Nervous System* (ed. D.R. Kankel & A. Ferrus), pp. 173–196. Elsevier–North Holland, New York.
1.3.6

Muntz W.R.A. (1963) The development of phototaxis in the frog (*Rana temporaria*). *Journal of Experimental Biology* **40,** 371–379.
3.1

Murphey R.K., Mendenhall B., Palka J. & Edwards J.S. (1975) Deafferentation slows the growth of specific dendrites of identified giant interneurones. *Journal of Comparative Neurology* **159,** 407–418.
1.2.1

Noirot E. (1964) Changes in responsiveness to young in the adult mouse: IV. The effect of an initial contact with a strong stimulus. *Animal Behaviour* **12,** 442–445.
4.2.1, 4.3.2

Noirot E. (1965) Changes in responsiveness to young in the adult mouse: III. The effect of immediately preceding performances. *Behaviour* **24,** 318–325.
4.2.1, 4.3.2

Norton-Griffiths M. (1967) Some ecological aspects of the feeding behaviour of the oystercatcher on the edible mussel. *Ibis* **109,** 412–424.
3.3.2

Norton-Griffiths M. (1968) The feeding behaviour of the oystercatcher (*Haematopus ostralegus*). D. Phil. thesis, University of Oxford.
3.3.2

Norton-Griffiths M. (1969) The organisation, control and development of parental feeding in the oystercatcher (*Haematopus ostralegus*). *Behaviour* **24,** 55–114.
3.3.2

Nottebohm F. (1969) The "critical period" for song learning in birds. *Ibis* **111,** 385–387.
3.4.1

Nottebohm F. (1970) Ontogeny of bird song. *Science* **167,** 950–956.
3.4

Nottebohm F. & Nottebohm M. (1978) Relationship between song repertoire and age in the canary *Serinus canarius. Zeitschrift für Tierpsychologie* **46,** 298–305.
3.4.1

Ohno S., Geller L.M. & Young Lai E.V. (1974) *Tfm* mutation and masculinization versus feminization of the mouse central nervous system. *Cell* **3,** 237–244.
1.3.7.

O'Keefe J. & Conway D.H. (1978) Hippocampal place units in the freely moving rat: why they fire when they fire. *Experimental Brain Research* **31,** 573–590.
5.4.3

O'Keefe J. & Nadel L. (1978) *The Hippocampus as a Cognitive Map.* Oxford University Press, Oxford.
5.4.3

Oppenheim R.W. (1981) Ontogenetic adaptations and retrogressive processes in the development of the nervous system and behaviour: a neuroembryological perspective. In: *Maturation and Development: Biological and Psychological Perspectives* (ed. K.J. Connolly & H.F.R. Prechtl), pp. 73–109. J.P. Lippincott, Philadelphia.
3.1

Oppenheim R.W. (1982) Preformation and epigenesis in the origins of the nervous system and behaviour: issues, concepts and their history. In: *Perspectives in Ethology, Vol. 5: Ontogeny* (ed. P.P.G. Bateson & P.H. Klopfer). Plenum Press, New York.
2.11

Owens N.W. (1975) A comparison of aggressive play and aggression in free-living baboons, *Papio anubis. Animal Behaviour* **23,** 757–765.
4.3.3

Oyama S. (1979) The concept of the sensitive period in developmental studies. *Merill-Palmer Quarterly* **25,** 83–103.
2.11

Oyama S. (1982) A reformulation of the idea of maturation. In: *Perspectives in Ethology Vol. 5: Ontogeny* (ed. P.P.G. Bateson & P.H. Klopfer). Plenum Press, New York.
2.11

Padilla S.G. (1935) Further studies on the delayed pecking of chicks. *Journal of Comparative Psychology* **20,** 413–443
2.7

Parkin D.T. (1979) *An Introduction to Evolutionary Genetics.* Edward Arnold, London.
1.6.1

Partridge L. (1978) Habitat selection. In: *Behavioural Ecology* (ed. J.R. Krebs & N.B. Davies). Blackwell Scientific Publications, Oxford.
1.6.4

Partridge L. (1981) Increased preferences for familiar foods in small mammals. *Animal Behaviour* **29,** 211–216.
3.3.2

Paulson G.W. (1965) Maturation of evoked responses in the duckling. *Experimental Neurology* **11,** 324–333.
2.8

Pavlov I.P. (1927) *Conditioned Reflexes.* Oxford University Press, Oxford.
5.3.1

Payne R.B. (1981a) Song learning and social interaction in indigo buntings. *Animal Behaviour* **29,** 688–697.
3.4.1

Payne R.B. (1981b) Population structure and social behavior: models for testing the social significance of song dialects in birds. In: *Natural Selection and Social Behavior* (ed. R.D. Alexander & D.W. Tinkle), pp. 108–120. Chiron Press, New York.
3.4.1

Payne R.B. (in press) Ecological consequences of song matching: breeding success and intraspecific song mimicry in indigo buntings. *Ecology*.
3.4.1

Pearce J.N., Colwill R.M. & Hall G. (1978) Instrumental conditioning of scratching in the rat. *Learning and Motivation* **9,** 255–271.
6.3.4

Pearce J.M. & Hall G. (1980) A model for Pavlovian learning: variations in the effectiveness of conditioned but not of unconditioned stimuli. *Psychological Review* **87,** 532–552.
5.3.2

Plomin R., de Fries J.C. & McClearn G.E. (1980) *Behavioral Genetics: a Primer*. W.H. Freeman, San Francisco.
1.1

Poindron D. & Le Neindre P. (1980) Endocrine and sensory regulation of maternal behavior in the ewe. *Advances in the Study of Behavior* **11,** 75–119.
4.2.1

Premack D. (1959) Towards empirical behavior laws: I. Positive reinforcement. *Psychological Review* **66,** 219–234.
6.3.3

Premack D. (1978) On the abstractness of human concepts: why it would be difficult to talk to a pigeon. In: *Cognitive Processes in Animal Behavior* (ed. S.H. Hulse, H. Fowler & W.K. Honig), pp. 423–451. Erlbaum, Hillsdale, New Jersey.
5.4.4

Quinn W.G. & Gould J.L. (1979) Nerves and genes. *Nature* (London) **278,** 19–23.
1.2.1, 1.5

Quinn W.G., Harris W.A. & Benzer S. (1974) Conditioned behavior in *Drosophila melanogaster*. *Proceedings of the National Academy of Sciences, USA* **71,** 708–712.
1.2.4

Quinn W.G., Sziber P.P. & Booker R. (1979) The *Drosophila* memory mutant *amnesiac*. *Nature* (London) **277,** 212–214.
1.2.4

Rabinowitch B. (1969) The role of experience in the development and retention of seed preferences in zebra finches. *Behaviour* **33,** 222–236.
3.3.2

Ramsay A.O. (1951) Familial recognition in domestic birds. *Auk* **68,** 1–16.
4.2.1

Ransom T.W. & Rowell T.E. (1973) Early social development of feral baboons. In: *Primate Socialization* (ed. F.E. Poirier), pp. 105–144. Random House, New York.
4.2.2

Ratner A.M. & Hoffman H.S. (1974) Evidence for a critical period for imprinting in khaki campbell ducklings (*Anas platyrhynchos domesticus*). *Animal Behaviour* **22,** 249–255.
2.8

Rescorla R.A. (1975) Pavlovian excitatory and inhibitory conditioning. In: *Handbook of Learning and Cognitive Processes Vol. 2.* (ed. W.K. Estes), pp. 7–35. Erlbaum, Hillside, New Jersey.
5.6

Rescorla R.A. & Wagner A.R. (1972) A theory of Pavlovian conditioning: variations

in the effectiveness of reinforcement and nonreinforcement. In: *Classical Conditioning II* (ed. A. Black & W.F. Prokasy), pp. 64–99. Appleton-Century-Crofts, New York.
5.3.2

Ressler R.H. (1963) Genotype-correlated parental influences in two strains of mice. *Journal of Comparative and Physiological Psychology* **56,** 882–886.
2.4

Restle F. (1957) Discrimination of cues in mazes: a resolution of the "place-vs-response" question. *Psychological Review* **64,** 217–228.
5.4.3

Restle F. (1958) Toward a quantitative description of learning set data. *Psychological Review* **65,** 77–91.
5.4.2

Revusky S. (1971) The role of interference in association over a delay. In: *Animal Memory* (ed. W.K. Honig & P.H.R. James), pp. 155–213. Academic Press, New York.
5.3.2, 6.4.1

Revusky S. (1977a) Learning as a general process with an emphasis on data from feeding experiments. In: *Food Aversion Learning* (ed. N.W. Milgram, L. Krames & T.M. Alloway), pp. 1–51. Plenum Press, New York.
6.3.1, 6.4.1

Revusky S. (1977b) Interference with progress by the scientific establishment: examples from flavor aversion learning. In: *Food Aversion Learning* (ed. N.W. Milgram, L. Krames & T.M. Alloway), pp. 53–71. Plenum Press, New York.
6.3.1

Ridley M. (1978) Paternal care. *Animal Behaviour* **26,** 904–932.
4.3.2, 4.5

Riechert S.E. (1978) Energy-based territoriality in populations of the desert spider *Agelenopsis aperta* (Gertsch). *Symposia of the Zoological Society of London* **42,** 211–222.
1.6.2

Riechert S.E. (1981) The consequences of being territorial: spiders, a case study. *American Naturalist* **117,** 871–892.
1.6.2

Roper T.J. (1973) Nesting material as a reinforcer in female mice. *Animal Behaviour* **21,** 733–740.
6.3.5

Roper T.J. (1975) Nest material and food as reinforcers for fixed-ratio responding in mice. *Learning and Motivation* **6,** 327–343.
6.3.5

Rosenblatt J.S. & Lehrman D.S. (1963) Maternal behavior of the laboratory rat. In: *Maternal Behavior in Mammals* (ed. H. Rheingold), pp. 8–57. John Wiley & Sons, New York.
4.2.1

Rothblat L.A. & Schwartz M.L. (1978) Altered early environment: effects on the brain and visual behavior. In: *Perception and Experience* (ed. R.D. Walk & H.L. Pick Jr), pp. 7–36. Plenum Press, New York.
3.2

Rothenbuhler W.C. (1967) Genetic and evolutionary considerations of social

behavior of honeybees and some related insects. In: *Behavior–Genetic Analysis* (ed. J. Hirsch), pp. 61–106. McGraw-Hill, New York.
2.7

Rozin P. & Kalat J.W. (1971) Specific hungers and poison avoidance as adaptive specialisations of learning. *Psychological Review* **78,** 459–486.
6.4.1, 6.5.1, 6.7

Rushforth N.B. (1965) Behavioural studies of the coelenterate *Hydra pirardi* Brien. *Animal Behaviour* Suppl. **1,** 30–42.
5.2.1

Rutter M. (1979) Maternal deprivation 1972–1978: new findings, new concepts, new approaches. *Child Development* **50,** 283–305.
4.2.4

Salzen E.A. & Meyer C.C. (1968) Reversibility of imprinting. *Journal of Comparative and Physiological Psychology* **66,** 269–275.
2.8

Schiller P.H. (1952) Innate constituents of complex responses in primates. *Psychological Review* **59,** 177–191.
5.4.1

Schilcher F. von & Hall J.C. (1979) Neural topography of courtship song in sex mosaics of *Drosophila melanogaster. Journal of Comparative Physiology* **129,** 85–95.
1.3.4

Schleidt W.M. (1961) Über die Auslösung der Flucht vor Raubvöbeln bei Truthühnern. *Naturwissenschaften* **48,** 141–142.
3.3.1

Schneiderman N. & Gormezano I. (1964) Conditioning of the nictating membrane of the rabbit as a function of CS–US interval. *Journal of Comparative and Physiological Psychology* **57,** 188–195.
5.3.1

Schneirla T.C. (1965) Aspects of stimulation and organization in approach/withdrawal processes underlying vertebrate behavioral development. *Advances in the Study of Behavior* **1,** 1–71.
3.3.1

Schneirla T.C. (1966) Behavioral development and comparative psychology. *Quarterly Review of Biology* **41,** 283–302.
2.2.1, 2.2.2

Schwartz B. (1978) *Psychology of Learning and Behavior.* Norton, New York.
5.6

Schwartz B. (1981) Autoshaping: driving toward a psychology of learning. *Contemporary Psychology* **26,** 823–825.
6.3.3

Seligman M.E.P. (1970) On the generality of the laws of learning. *Psychological Review* **77,** 406–418.
6.2.4, 6.3, 6.7

Seligman M.E.P. & Hager J.L. (eds) (1972) *Biological Boundaries of Learning.* Prentice-Hall, Englewood Cliffs, New Jersey.
6.3, 6.7

Sevenster P. (1973) Incompatibility of response and reward. In: *Constraints on Learning: Limitations and Predispositions* (ed. R.A. Hinde & J. Stevenson-Hinde), pp. 265–283. Academic Press, London.
6.3.3

Shettleworth, S.J. (1972) Constraints on learning. *Advances in the Study of Behavior* **4,** 1–68.
6.2.4, 6.3, 6.4.1, 6.6.2, 6.7

Shettleworth S.J. (1975) Reinforcement and the organisation of behavior in golden hamsters: hunger, environment and food reinforcement. *Journal of Experimental Psychology: Animal Behavior Processes* **1,** 56–87.
6.3.4

Shettleworth S.J. (1978) Reinforcement and the organisation of behavior in golden hamsters: sunflower seed and nest paper reinforcers. *Animal Learning and Behavior* **6,** 352'–362.
6.3.4

Shettleworth S.J. (1981) Reinforcement and the organisation of behaviour in golden hamsters: differential overshadowing of a CS by different responses. *Quarterly Journal of Experimental Psychology* **33B,** 241–256.
6.3.4

Shettleworth S.J. (in press) Function and mechanism in learning. In *Advances in Analysis of Behavior Vol. 3: Biological Factors in Learning* (ed. M. Zeiler & P. Harzem). John Wiley & Sons, New York.
6.6.3

Shettleworth S.J. & Juergensen M.R. (1980) Reinforcement and the organisation of behavior in golden hamsters: brain stimulation reinforcement for seven action patterns. *Journal of Experimental Psychology: Animal Behavior Processes* **6,** 352–375.
6.3.4

Shlaer S. (1971) Shift in binocular disparity causes compensatory change in the cortical structure of kittens. *Science* **173,** 638–641.
3.2

Siegel R.W. & Hall J.C. (1979) Conditioned responses of courtship behavior of normal and mutant *Drosophila. Proceedings of the National Academy of Sciences, USA* **76,** 3430–3434.
1.2.4

Skinner B.F. (1938) *The Behavior of Organisms.* Appleton-Century-Crofts, New York.
5.3.1

Slater P.J.B. (in press) Bird song learning: theme and variations. In: *Perspectives in Ornithology* (ed. A.H. Brush & G.A. Clark Jr). Cambridge University Press, New York.
3.4, 3.4.4, 3.6

Slater P.J.B. & Ince S.A. (1979) Cultural evolution in chaffinch song. *Behaviour* **71,** 146–166.
3.4.3

Slater P.J.B. & Ince S.A. (1982) Song development in chaffinches: what is learnt and when? *Ibis* **124,** 21–26.
3.4, 3.4.1, 3.4.3

Slater P.J.B., Ince S.A. & Colgan P.W. (1980) Chaffinch song types: their frequencies in the population and distribution between the repertoires of different individuals. *Behaviour* **75,** 207–218.
3.4.3

Sluckin W. (1972) *Imprinting and Early Learning.* 2nd edn. Methuen, London.
4.2.1

Small W.S. (1901) Experimental study of the mental processes of the rat. *American Journal of Psychology* **12,** 206–239.
5.4.3

Smith F.V. & Nott K.H. (1970) The "critical period" in relation to the strength of the stimulus. *Zeitschrift für Tierpsychologie* **27,** 108–115.
2.8

Smith J.C. & Roll D.L. (1967) Trace conditioning with X-rays as the aversive stimulus. *Psychonomic Science* **9,** 11–12.
6.3.1

Smith P.K. (1982) Does play matter? Functional and evolutionary aspects of animal and human play. *Behavioral and Brain Sciences* **5.**
4.3.3, 4.5

Snow C. (1972) Mother's speech to children learning language. *Child Development* **32,** 541–564.
4.2.1

Sotelo C. (1980) Mutant mice and the formation of cerebellar circuitry. *Trends in Neurosciences* February, 33–36.
1.3.6

Staddon J.E.R. & Simmelhag V.L. (1971) The "superstition" experiment: a reexamination of its implications for the principles of adaptive behavior. *Psychological Review* **78,** 3–43.
6.3, 6.3.3

Stent G.S. (1980) The genetic approach to developmental neurobiology. *Trends in Neurosciences* February, 49–51.
1.5

Stent G.S. (1981) Strength and weakness of the genetic approach to the development of the nervous system. *Annual Review of Neurosciences* **4,** 163–194.
1.2.1, 1.2.3, 1.5, 1.7

Sternglanz S., Gray J.L. & Murakami M. (1977) Adult preferences for infantile facial features: an ethological approach. *Animal Behaviour* **25,** 108–115.
4.2.1

Stevenson-Hinde J. (1973) Constraints on reinforcement. In: *Constraints on Learning: Limitations and Predispositions* (ed. R.A. Hinde & J. Stevenson-Hinde), pp. 285–299. Academic Press, London.
6.3.5

Stewart P.A. (1958) Local movements of wood ducks (*Aix sponsa*). *Auk* **75,** 157–168.
4.2.1

Stewart P.A. (1974) Mother wood ducks feeding away from their broods. *Bird Banding* **45,** 58.
4.2.1

Stryker M.P. & Sherk H. (1975) Modification of cortical orientation selectivity in the cat by restricted visual experience. *Science* **190,** 904–906.
3.2

Sulston J.E. & Horvitz H.R. (1977) Postembryonic cell lineages of the nematode *Caenorhabditis elegans*. *Developmental Biology* **56,** 110–156.
1.2.2

Sulston J.E. & White J.G. (1980) Regulation and cell autonomy during postembryonic development of *Caenorhabditis elegans*. *Developmental Biology* **78,** 577–597.
1.2.2

Sulston J.E., Albertson D.G. & Thomson J.N. (1980) The *Caenorhabditis elegans* male: postembryonic development of non-gonadal structures. *Developmental Biology* **78,** 542–576.
1.2.2

Suzuki S., Augerinos G. & Black A.H. (1980) Stimulus control of spatial behavior on the eight-arm maze in rats. *Learning and Motivation* **11,** 1–18.
5.4.3

Thorndike E.L. (1911) *Animal Intelligence.* Macmillan, New York.
5.3.1, 6.3.4

Thorpe W.H. (1956) *Learning and Instinct in Animals.* Methuen, London.
2.3

Thorpe W.H. (1958) The learning of song patterns by birds, with especial reference to the song of the chaffinch *Fringilla coelebs. Ibis* **100,** 535–570.
3.4

Tinbergen N. (1951) *The Study of Instinct.* Oxford University Press, London.
6.5.2

Tinbergen N. (1953) Specialists in nest-building. *Country Life* (January), 270–271.
2.3

Tinbergen N. & Perdeck A.C. (1950) On the stimulus situation releasing the begging response of the newly hatched herring gull chick (*Larus argentatus argentatus* Pont.). *Behaviour* **3,** 1–39.
3.2

Trevarthen C. (1975) Early attempts at speech. In: *Child Alive* (ed. R. Lewin), pp. 62–80. Temple Smith, London.
4.2.1

Tryon R.C. (1940) Genetic differences in maze-learning ability in rats. *Yearbook of the National Society for the Study of Education* **39,** 111–119.
Introduction

Vauclair J. & Bateson P.P.G. (1975) Prior exposure to light and pecking accuracy in chicks. *Behaviour* **52,** 196–201.
2.7

Vieth W., Curio E. & Ernst U. (1980) The adaptive significance of avian mobbing. III. Cultural transmission of enemy recognition in blackbirds: cross-species tutoring and properties of learning. *Animal Behaviour* **28,** 1217–1229.
3.3.1

Waddington C.H. (1935) *How Animals Develop.* Allen & Unwin, London.
Introduction

Wagner A.R. (1976) Priming in STM: an information-processing mechanism for self-generated or retrieval-generated depression in performance. In: *Habituation: Perspectives from Child Development, Animal Behavior, and Neurophysiology* (ed. T.J. Tighe & R.N. Leaton), pp. 95–128. Erlbaum, Hillsdale, New Jersey.
5.2.1

Wagner A.R., Logan F.A. Haberlandt K. & Price T. (1968) Stimulus selection in animal discrimination learning. *Journal of Experimental Psychology* **76,** 171–180.
5.3.2

Walk R.D. & Pick H.L. Jr (1978) *Perception and Experience.* Plenum Press, New York.
3.6

Ward S. (1973) Chemotaxis by the nematode *Caenorhabditis elegans:* identification of attractants and analysis of the response by use of mutants. *Proceedings of the National Academy of Sciences, USA* **70,** 817–82.
1.2.2

Ward S. (1977) Invertebrate neurogenetics. *Annual Review of Genetics* **11,** 415–450.
1.2.2

Ward S., Thomson J.N., White J.G. & Brenner S. (1975) Electron microscopical reconstruction of the anterior sensory anatomy of the nematode *Caenorhabditis elegans. Journal of Comparative Neurology* **160,** 313–338.
1.2.2

Watson J.B. (1930) *Behaviorism,* 2nd edn. Norton, New York.
Introduction, 6.2.4

Wells M.J. (1978) *Physiology of an Advanced Invertebrate.* Chapman & Hall, London.
5.2.2

Wheeler W.M. (1910) *Ants: their Structure, Development and Behavior.* Columbia University Press, New York.
2.6

White J.G., Southgate E., Thomson J.N. & Brenner S. (1976) The structure of the ventral nerve cord of *Caenorhabditis elegans. Philosophical Transactions of the Royal Society of London* B **275,** 327–348.
1.2.2

Wiesel T.N. & Hubel D.H. (1974) Ordered arrangements of orientation columns in monkeys lacking visual experience. *Journal of Comparative Neurology* **158,** 307–318.
3.2

Wilcoxon H.C., Dragoin W.B. & Kral P.A. (1971) Illness-induced aversions in rat and quail: relative salience of visual and gustatory cues. *Science* **171,** 826–828.
6.3.1, 6.3.2

Williams B.A. (1978) Information effects on the response–reinforcer association. *Animal Learning and Behavior* **6,** 371–379.
5.3.2

Williams D. & Williams H. (1969) Automaintenance in the pigeon: sustained pecking despite contingent non-reinforcement. *Journal of the Experimental Analysis of Behavior* **12,** 511–520.
6.3.3

Willmund R. & Ewing A. (1982) Visual signals in the courtship of *Drosophila melanogaster. Animal Behaviour* **30,** 209–215.
1.1

Wilson E.O. (1971) *The Insect Societies.* Harvard University Press, Cambridge, Mass.
2.6

Wilson E.O. (1978) *On Human Nature.* Harvard University Press, Cambridge, Mass.
Introduction

Woolfenden G.E. (1975) Florida scrub jay helpers at the nest. *Auk* **92,** 1–15.
4.3.2

Zentall T. & Hogan D. (1974) Abstract concept learning in the pigeon. *Journal of Experimental Psychology* **102,** 393–398.
5.4.4

INDEX